CONFESSIONS OF AN ECCENTRIC DREAMER

CONFESSIONS OF AN ECCENTRIC DREAMER

An Autobiography

Peter James McLean
With
Carolyn Lee (McLean) Bosetti

CONFESSIONS OF AN ECCENTRIC DREAMER
AN AUTOBIOGRAPHY

Copyright © 2016 Peter James McLean and Carolyn Lee (McLean) Bosetti.

All rights reserved. No part of this book may be used or reproduced by any means, graphic, electronic, or mechanical, including photocopying, recording, taping or by any information storage retrieval system without the written permission of the publisher except in the case of brief quotations embodied in critical articles and reviews.

iUniverse books may be ordered through booksellers or by contacting:

iUniverse
1663 Liberty Drive
Bloomington, IN 47403
www.iuniverse.com
1-800-Authors (1-800-288-4677)

Because of the dynamic nature of the Internet, any web addresses or links contained in this book may have changed since publication and may no longer be valid. The views expressed in this work are solely those of the author and do not necessarily reflect the views of the publisher, and the publisher hereby disclaims any responsibility for them.

Any people depicted in stock imagery provided by Thinkstock are models, and such images are being used for illustrative purposes only.
Certain stock imagery © Thinkstock.

ISBN: 978-1-4917-5175-6 (sc)
ISBN: 978-1-4917-5177-0 (hc)
ISBN: 978-1-4917-5178-7 (e)

Library of Congress Control Number: 2014919557

Printed in the United States of America.

iUniverse rev. date: 02/20/2016

TABLE OF CONTENTS

Preface xi
The "Runaway" 1
In the Beginning 14
Getting Started 27
The Early Years in Windsor 41
Inventions Brought to Life 50
The Boy's Club 63
Treasure Hunting 78
Not My Finest Moments 96
Boating for Dummies 108
Accidental Discovery 121
Bancroft Adventures 130
Wawa Adventures 156
More Adventures 186
Troubled Waters 192
Home Again – St. Thomas 204
The Itch Is Back 210
Marion and Me in Later Years 225
The End Is Near 235
Epilogue 239
Timeline Reference 241

This book is dedicated to Marion, my wife of sixty-eight years, who passed away in January, 2007 and to my children.

ACKNOWLEDGEMENTS

I want to thank Richard Strong and Robert Woodhouse for reading the original manuscript and offering suggestions and also thanks to Marnie Pouget, constituency assistant to Jeff Watson, MP for Essex, and Ron Tozer, Algonquin Park Archivist (volunteer).

    I need to express my special thanks to my two daughters. I have been very blessed having Barbara as our own private nurse and Carolyn who helped me write this book. Without their help, this book would not have been written.

Peter James McLean

I want to thank my son, Kai Richard Bosetti, my brother, James Guy McLean, Richard Strong, Karen Hughes, Margaret Dowling/Hassett, Mike Todd, Carolyn and Kim Kehoe and Dr. E. J. Morris as well as Mickey and Dolly Clement for their help. I also want to thank the publishing staff at iUniverse for their invaluable assistance.

Carolyn (McLean) Bosetti

PREFACE

For years people have been telling me that my life reads like fiction. My family and friends know differently, but few others can really believe the truth of it.

I have met many interesting people and have done many things in my life. I believe that everyone should follow their dreams.

Many won't believe that someone would risk it all, time and time again, for no other reason than just to see if something can be done. But I did, and if given the chance, I would do it again. We all have our good times and bad times. These are natural and don't make us any better or worse than anyone else. They just set us apart.

This is the story of *me*. As with every story, there are many chapters. I found the key to my story early in life. When one chapter came to its logical conclusion, I let it end. I don't believe in looking back with regret; it's a waste of time.

I hope that with this book I can inspire those would-be adventurers to go out and live. The words "cannot be done" and "impossible" aren't in my vocabulary.

Remember, all the stories in this book are true but some names have been changed.

I hope you enjoy reading these stories as much as I enjoyed telling them.

Peter James McLean

Living with my father was always interesting. I knew he worked for the railroad but knew only bits and pieces about his other adventures.

Peter James McLean

There appeared to be lots of ups and downs in our home life, but they were never openly discussed.

I went through my early years never knowing people had to live on a regular paycheck. Money was always there when something was needed, so I never had to question where it came from.

Dad and I wrote a draft of this book in 2007 and had copies printed. After reviewing what had been written, I realized that he had left out or glossed over the more colorful episodes in his life. When he finally talked me into helping him finish this book we agreed that he had to air all his laundry; good and bad.

As I learned about my father through all this, it was, at times, hard to digest. As a daughter, I only wanted to see my father as a do-no-wrong person. Oh, I knew there were problems growing up because of his alcoholism, but I guess I just chose to ignore them.

Also, through the writing of this book, I learned about the man my Dad really was and found him more interesting and complicated than I would ever have guessed.

Dad passed away on June 26, 2010, at the age of ninety-two. He was every bit the eccentric dreamer.

I made him a promise that this book would be published as he wanted it to be, and I have kept my promise.

Love you, Dad!

Carolyn (McLean) Bosetti

THE "RUNAWAY"

During my second year in vocational school, Cliff Palmer and I were walking on Flora Street, and we decided to run away that afternoon. It was a warm, bright, and sunny day in May of 1934. I was sixteen years old.

It was too nice a day for us to be sitting in a classroom. So we tore up all our schoolbooks, threw them away, and then headed out. It took us two days to get to Huntsville, Ontario. Both of us were tired and hungry, and we had no money. Cliff decided he couldn't continue, and he went back home.

I wasn't about to give up that easily and decided to keep on going. My goal was to get a job on a freighter out of Halifax, Nova Scotia, and sail on the high seas.

While in Huntsville, I was associating with the hobos who were riding the rails and was excited by the prospect of riding with them. I eagerly joined the group, and we rode the freight trains to Montreal, Quebec.

After we arrived there, I went into a combination beauty parlor and barbershop to ask for money for something to eat. The lady who owned the establishment asked me if I would clean the garbage out of the alley behind the shop in exchange for food and a place to sleep in a small backroom. I took her up on her offer and thus had someplace to stay while there.

During that time, I met someone who ran different amusement games on what the hobos referred to as Little Acre at the foot of Saint Lawrence Main at the Canals. He gave me a job running the Crown and Anchor Board. I did that for about two weeks and was surprised at how much money would change hands. All the hobos went there with all the money they had bummed just to gamble. I earned enough money to join the Sailor's Institute in Montreal.

From Montreal, I rode the freight trains to Halifax, Nova Scotia, and stayed at the Sailor's Institute there since I was now a member. I felt sure I would be hired to work on one of the ocean-going ships, but they kept telling me I was too young and inexperienced and that they had no job for me. This put an end to my dream, and I had to move on—but to what?

It was back to the rails!

A blind tender is the open car directly behind the engine where the coal needed to run the steam engine was kept. Naturally, it was extremely dirty. The first train out of Halifax was a passenger train heading for Moncton, New Brunswick, so I joined some other hobos and hopped on the blind tender. As we were pulling into Moncton, the alarm went out that the railroad police were on each side of the depot, waiting to arrest the hobos.

As the train slowed, I jumped off, skidded, and rolled in the cinders along the side of the train tracks. I had been riding the train for so long that I misjudged how fast we were traveling. My hands and the side of my face were a terrible mess from all the cinders. Luckily I didn't go under the wheels of the train or break any bones.

I prowled around Moncton in a residential district trying to find something to eat and a place to get cleaned up. I was filthy from the cinders and from riding the blind tender. One house had a light in the front window and looked so inviting I knocked on the door. The lady who answered felt sorry for me because I was such a mess. She wanted to know what happened, and so I told her about running away and how I ended up at her door. She said that if I was willing to clean up her yard, I'd be able to stay there and eat for a couple of days until I healed up.

The lady even insisted on removing all the cinders from my face and arms. In fact, she took such good care of me that I healed very quickly.

Instead of being there only for a couple of days, I was there for an entire week. Her backyard was a mess of debris and firewood. I piled up all the firewood at the side of her house and cleaned and raked the lawn and was even able to repair her lawnmower.

I thanked her and expressed my appreciation for all that she had done for me. Since she was an elderly widow, I got the impression she was glad to have someone to take care of, even if it was only for a week. She also said she was happy with all the work that I had done for her.

I was well fed and feeling pretty good, and thoughts of a new adventure got to me again. I figured I would leave and take a freight train back to Montreal—which I did.

When I arrived back in Montreal, I immediately went down to the Canals to check things out. I was thinking I could get my job back running the Crown and Anchor Board. In the short time I had been away, the police had cleared out all the bums and hobos, and the gambling games were gone. It seemed that for me Montreal had changed for the worse, and I didn't see any reason to stick around.

After prowling around for a while, I found myself in a residential neighborhood that I felt would be a good area where I could scrounge something to eat from one of the homes.

I knocked on the back door of one house that looked to be occupied by people of means. An older fat lady answered the door, and I told her what I wanted. She started growling and swearing at me in French. Although I didn't speak French, it was easy to understand the meaning of her words. She turned around and cut a chunk of stale bread and tossed it at me. It made me so mad that I caught the piece of bread and threw it right back at her. It hit her in the head and she fell into a corner of the kitchen and landed on the floor. She then proceeded to scream at the top of her lungs and I took off running through back yards and hopping over fences.

All the yards were fenced, and the fences were made of wood. As soon as I put my weight on a fence to hop over it, the wood would break. This happened in yard after yard. It started all the dogs in the area barking and had the neighbors spilling out of their houses to see what the commotion was all about. I kept running through the yards for about two blocks, and I knew by the uproar I was causing that the police would be coming soon. Finally, I got to a street where I could see train tracks. There was a freight train just starting to pull out of the yard, and I jumped on it. I figured that I'd better get out of town in a hurry.

When I had a chance to look to see what type of cars made up the train, it turned out to be all hopper cars, and the cars were all filled with coal. The only place to ride was down in the corner of the coal car, where you were out of the wind. I stayed on that coal train till the next morning when they pulled into the yard in Ottawa, Ontario, the capital of Canada. When I got off the train, I was covered in coal dust from top to bottom and I was blacker than the ace of spades. You could barely see the whites of my eyes because the wind blew soot all over me during the trip.

At the first gas station I came to, I was thinking about getting washed up a bit when the owner came out the door hollering and telling me to get the hell out of there. He said I was too dirty to be washing up in his gas station, that I had no business being there, and to get out and stay out.

I went outside and was standing there, trying to think of what to do next, when the owner came over to me and said he would help me out by putting a cover over the seat of his truck and then driving me to the Salvation Army Hostel so I could go in and ask for help. I told the man that I really appreciated his help. At the hostel they supplied me with clean clothes after I finally got cleaned up.

The commander of the Salvation Army seemed to have some sympathy for me. He wanted to know where I was going and if I had a job or could I get one. I told him there wasn't any job I was qualified for. He told me that the government was running make work camps similar to the CCC camps in the United States. The name stood for Civilian Conservation Corps camps, and they were created during the Depression to provide employment for the jobless.

The commander said he knew the manager of the main office there in Ottawa and that they had camps for unemployed people all over Canada. He thought he might be able to get me a job in one of these camps. Until he told me about them, I had never heard anything about them.

I didn't particularly care where I went as long as I got a job. As he was going to check things out to see what he could do for me, he asked me if I had any identification. I told him that all I had was my certificate from the Sailor's Institute in Montreal and Halifax, and it showed that I was twenty-one years old. It turned out the Camps wouldn't hire anyone under twenty-one.

The commander then took me to meet this manager. When the manager asked, I told him I preferred to work in Ontario, but after checking his files he told me the camps in Ontario were all full. The Salvation Army commander asked him to please check again, and they finally found one with an opening. That opening was in Camp Seven where they were building Highway 60 through Algonquin Park. They told me that if I accepted the job I would be given a train ticket to get there.

The following day, the commander gave me a train ticket and three government meal vouchers, and I was on my way to Camp Seven. I didn't have to ride the rails as a hobo this time. There I was, sitting in a coach, riding the rails like a king. I kept thinking to myself how lucky I was because I had really hit pay dirt in Ottawa.

Camp Seven

When I got to the camp, I reported to the superintendent, who said he had been expecting me and that I had a job there.

I started working at Camp Seven the first part of October 1934. The superintendent explained that my wages would be thirty dollars a month, including room and board. However, I had to buy my own clothes and any personal needs from the commissary. Any belongings that I needed to get started I could charge and have the cost deducted from my wages at a later date. And, because the commissary was government sponsored and not for profit, the price of everything they carried was relatively low.

I was such a greenhorn that the superintendent didn't know what to do with me, but he felt if he made me the water boy, mailman, and errand boy to start with, he could figure out later what job in the camp I would be best suited for. The superintendent wasn't under a lot of pressure and just wanted the camp to run smoothly.

As the water boy, it was my responsibility to provide fresh water for the men working in the bush. I used a single horse and wagon and had to make the complete rounds to all the different men who had to be taken care of. The jobs they were doing ranged from cutting down trees and using horses to clear the stumps to drilling holes for the dynamite man to blow the rough cuts while others were using horses to scrape and level the roadway.

I enjoyed my work as well as all the other odd jobs I was given, because it allowed me to circulate and meet and talk to all the different crews and workers in the camp. Another job I had was to take the mail out to the Highland Inn, which served as a tourist hotel, post office, and railroad station, all in one complex.

Before coming here, I didn't realize how much manual labor it took to build a road. They did not have any mechanical equipment in the camp. There were no bulldozers, no chain saws, and no pneumatic drills for drilling holes in the rocks and stumps. Everything was done manually with only horses for help. We even had our own blacksmith to sharpen drill bits and axes and to do all the necessary metal work.

All the holes in the rocks for the dynamite were drilled by hand. It took four men to drill each hole. They had one man sitting in the middle holding the steel drill bit, and then three men would stand around him in a circle. Each man would swing his sledgehammer and hit the top of the drill bit in rhythmic blows.

It was great to watch them because you kept thinking about what would happen if one of the boys missed the drill. That never happened while I was there. The boys did this for eight hours a day. It amazed me how fast they could drill the dynamite holes through solid rock. They had arm muscles that would out shine body builders.

The dynamite man would pack the dynamite sticks in the holes, attach the cap on the end of the fuse, and cut the fuse a certain length according to the time that he needed to clear the areas before he set off the charge.

One of the biggest problems in the camp was getting rid of the tree stumps because of their roots. The horses were unable to pull out large stumps, so they had to be dynamited out of the ground separately.

All the trees were cut down using axes and crosscut saws. When cutting down the trees, the men had to use their expertise to have the tree fall exactly in an open area so that the tree wouldn't fall against another tree and be left hanging in the air. They were able to cut and fell a tree where it was needed.

A short time after I started working at the camp, there was an opening for a "powder monkey." The job consisted of helping the dynamite expert. To do the job, you rammed a heavy bar into the ground around the stump to make holes about three feet deep, and then you placed two to three sticks of dynamite in the holes and put on the caps to get them ready to light. You would cut the first fuse about an inch long and then the next fuse one inch longer than the previous one until all the fuses were lit. This was necessary so that the dynamite would blow in succession. Using an acetylene torch with a big flame, you would run from one fuse to the next, lighting them, and you then kept on running until they were all lit and you were in the clear, usually behind a large tree. In order to be certain that all the dynamite exploded, you had to count the explosions.

If one didn't explode, you had to make a note of it and not allow anyone to walk in the proximity of that stump. For safety reasons, you could not check any fuses that didn't blow until the next day, because the fuse could just be smoldering and could still blow. Then you would go back and check to see why it didn't blow. You then had to refuse the dynamite with the caps and blow them up again. The caps for the fuses are more dangerous than the dynamite because they blow first and that sets off the dynamite.

I was happy to get this job and with practice I could light 12 fuses for 12 stumps at a time and they would go off exactly in the rotation they should. The wood from the stumps was used as fire wood to keep the men warm while they ate their lunch.

I quickly learned that dynamite blows down, not up, as most people think. In order to shatter and break up large chunks of rocks, I would have to put four or five sticks of dynamite on top of the rock on a flat area and then pour half a pail of mud over the dynamite. The mud held the dynamite in place while I put on the cap, lit the fuse, and got in the clear. The force of the explosion went downward and shattered the rock.

In the camp, eight men were assigned to each bunkhouse, and you would be assigned to either an upper or lower bunk. There was a large wood stove at one end to provide heat. Everyone ate breakfast and dinner in the dining hall, which was located in a separate building.

Lunch would be brought to the crews on a lunch wagon so they could be served a hot lunch at the various work sites.

In the bunkhouse I was assigned to, there was a fellow named Danny Barton, who was from Orillia, Ontario. Danny and I got to be very friendly, and we played checkers every chance we got. We got so good at the game that the first two moves on the checkerboard would determine the winner.

Every Friday night after supper, a poker game started, and it ran continuously until it was time to go back to work on Monday morning. The poker game was mostly five and seven-card stud with a five-to-ten-cent limit. Anybody in the camp was welcome to sit at the table and join in the game. You could come into the game or leave it whenever you wanted. It was one of the few recreational things we had at the camp.

Another person in my bunkhouse was named Gene Tunney, and the reason everyone knew his name was because Gene Tunney was the name of the boxing world champion at the time.

This Gene was not interested in boxing, but he was a great poker player and the luckiest person I'd ever seen play the game. He made enough money from his poker playing to send some home to his mother in Orillia. His mother was totally dependent financially on Gene.

When I first got to Camp Seven, the men showed me how to take a Swedish steam bath. It was in a separate building, and there were rocks piled around a lit fire pit. The rocks would get red-hot. You would take a pail of water and throw it on the rocks, and hot steam would fill the room. You would go in and sit there with a bar of soap, and as you started sweating, you would wash yourself down.

The first two or three steam baths I took were much needed just to get all the coal dust off me. At the end of the bath, the first person to come out would wait for the next person and then douse them with ice water. You would gasp when the water first hit you, but then you could run back to your bunkhouse in the snow without feeling the cold. You could take a steam bath in both summer and winter.

When I first arrived at the camp, the superintendent told me that there were two strict rules for all the workers: no fighting and no stealing. Anyone involved would be discharged immediately, and it was enforced by a park ranger. We were reminded that Algonquin Park was a tourist destination for recreational activities and that it was also an animal sanctuary.

The park ranger was an expert axe man. At different times he would show me how accurate he really was. For example, he would pick out a tree that was leaning in a certain direction and tell me where the tree would land. He would chop the tree down, and as it was falling, it would swing around 180 degrees and fall on the ground within a foot of where he said it was going to fall. He could pick out a tree twenty-five to thirty feet away and would throw the axe and embed it right into the tree he had picked out.

One night the superintendent called me and told me to go down to the horse stables and see what was going on. The horses were raising quite a commotion, kicking their stalls and jumping around. He wanted to know exactly what was happening.

When I got to the stable, low and behold, there was a car sitting there. It was the first car I had seen at the camp. I hadn't thought the road coming into the camp was fixed up enough for a car to travel over it, but I guess I was wrong.

When I went into the stables, there was a crowd of men from the camp in there, including Danny Barton, my checkers partner.

I asked Danny, "What the hell is going on? The superintendent just sent me to find out what's happening in the stables to upset the horses so much and he wants me to report back to him promptly."

Danny said he had really hit pay dirt! He said he had four female residents who were on leave from the mental health facility in Orillia up in the hayloft. The men paid Danny a certain amount of money so they could go up in the hayloft with the girls. He had a long line of men waiting at the ladder leading to the hayloft. The other men who had already been up there were bragging about what kind of time they had and agreeing it was worth the money.

I again told Danny that the superintendent was waiting to hear what was causing the ruckus. I told him he had better pack up the women and take off. Danny loaded the women in the car—which, by the way, was stolen—and took off.

I went back to the superintendent and told him what was going on but that the women just left in a car. He wanted to know who was responsible, but I told him that there were just too many men there for me to know. He grabbed the telephone and called Camp Six, who said the car just went by there at a high rate of speed. He proceeded to call all the other camps to stop the car, but they didn't get the message soon enough, and Danny got the girls out of the camp without being caught.

At the time, Danny was supposed to be at home on sick leave. He thought taking the girls to the camp was a foolproof way to earn extra money. After he got back to the camp when his sick leave was over, he said I had saved his life by not mentioning his name to the superintendent. He said he had gotten the women out of the camp and booked back into the mental health facility as fast as possible. He also got the car back where it belonged before it was reported stolen.

The thinking at that time was that the girls from the health unit got out at certain times for short leaves, and the men in Orillia felt the women were fair game for anyone. Everybody involved with them was happy because the men knew the women were all medically clean and healthy.

I was a spectator to only one fight in the camp, and I can't remember the names of the two men involved. One was a Frenchman from Quebec who would brag about how he could flip upside down in the bunkhouse and put his footprints on the ceiling. He was always showing everyone how light he was on his feet. Also staying in my bunkhouse was a big, well-built man who was always telling humorous stories about the different altercations he had been in and how he had always won. For about two weeks straight, the Frenchman kept nagging the big guy and saying how he could easily beat him because he was so fast and quick on his feet.

One day we were sitting by a big fire to keep warm while we ate our lunch when the teasing started again. This time, however, the big buy told the Frenchman he was going to end it right then and there. He grabbed the Frenchman, who weighed 180 pounds, by the scruff of the neck and one leg, upset him, and held him bum side down over the fire. He held him there for well over five minutes with the flames burning the Frenchman's pants. The Frenchman was screaming and trying to fight the big guy, but it had no effect. The big guy just kept holding him until his pants burned off. Then the big guy threw the Frenchman off to one side, into the snow. He told him that that should keep him from nagging him anymore and take the chip off his shoulder.

The Frenchman had severe burns on his bum and backside and couldn't work or sit down. It was necessary to have the first aid man look at him, and he said the man had to go to the hospital because of the seriousness of his burns. He was taken away on a stretcher and shipped by train to a hospital.

The park ranger investigated the incident and found out how it happened. The Frenchman was terminated immediately, but, the big guy was exonerated. It was determined he was not responsible for any portion of the uproar because of his good nature. Holding 180 pounds straight out over the fire as he had done showed him to be an extremely powerful individual. It made you realize that all the good-natured talks of his fighting were true.

Another man in our bunkhouse had Parkinson's disease, but everyone at that time called it "Saint Vitus' dance" because of all the involuntary movements. He had come into the camp before I did, and everybody teased him mercilessly. Not understanding what was wrong with the fellow, I went right along with the teasing. One day the man decided he couldn't take the bullying anymore, and he went into the forest and shot himself.

When I heard this, I was racked with guilt at having been so cruel to him. Since that day I have never gotten involved in any type of teasing or bullying of someone who had a disability of any kind. Whenever I hear about kids being bullied at school who turn violent toward their schoolmates, I think of that man in Camp Seven. He could have just as easily started shooting the men who teased him. Instead he chose to commit suicide.

During my time at the camp, I learned a lot of things and met a lot of very nice, hardworking people. One of the things that come to mind was that when you could hear a tree cracking in the bush in the wintertime, you knew it was thirty-plus degrees below zero. At this temperature, the tree sap explodes just like gunpowder. Another thing I was shown was how to wrap your feet in newspapers before you put your boots on. It would keep your feet warm and dry because at different times in the winter we didn't have socks available to wear.

By the following spring, I had money in my pocket, and during the month of May I got spring fever, quit my job, and decided to head home.

Peter James McLean

I recently found a photograph taken of my fellow workers and me at the camp. It is amazing that these pictures still exist. I can't remember who took them or why they were taken or how I ended up with them.

My mother actually saved two letters I had written to her in 1935 while I was at Camp Seven, and in later years, she gave them to me. I still have them.

Top: Gene Tunney and myself (on right) 1934
Center: Camp Seven during the time I was working there
Bottom: Myself and co-workers

Home Again

By the time I got home, I was in as sorry a shape as you can imagine. I looked like all the hobos you ever saw and read about. My mother nearly had a fit when she saw me and made sure I got cleaned up and fed properly. In fact, she had to cut the socks off my feet, because I hadn't been able to wash my feet for at least three months.

Now that I was home, I couldn't find a full-time job, so I ended up working several part time jobs, ranging from hoeing tobacco to serving beer at a hotel in Port Stanley and unloading lake freighters. I also delivered ice and coal for Tommy Johnstone, who had married my sister, June. They had a good business. In the beginning, Tommy would buy his ice from an icehouse in Port Stanley, but a short time later, he had his own icehouse on Wellington Street in St. Thomas.

Young people today can't fathom having an icebox, which served as the household refrigerator, and having ice delivered to their homes or apartments. The ice was made in three-hundred-pound units that had to be chopped up using an ice pick into twenty-five, fifty or one-hundred-pound blocks. Most iceboxes held at least fifty-pound blocks of ice. Using large ice tongs, sometimes you had to walk up three flights of stairs carrying the ice, so you had to be in good shape.

I did these odd jobs until I finally got a full-time job.

Cleaned up and ready to get a job

IN THE BEGINNING

I was born January 3, 1918, the fourth of seven children of James Guy and Charlotte Loftus McLean. My grandfather's family came to Canada from Scotland in 1816. When Dad's father died, he left farmland to each of his children. My dad, however, didn't want to farm, so he used his inheritance to attend Queen's University to study mining engineering. The McLean's were always proud of their link with the McLean clan of Mull. Dad was also a member of the Masonic Lodge and had a fifty-year jewel.

Mother and Dad met in the St. Thomas/Fingal area, and they married in October 1911 in Sudbury, Ontario, where they started out their married life together in a log cabin in Northern Ontario at Crerar. Dad worked for the Canadian Copper Company in Coppercliffe and Cobalt, Ontario. He also did some prospecting during that time and staked a gold mining claim on the property where their log cabin was located. It consisted of forty acres. Dad ended up buying the property and never sold it.

In early 1916, Mother inherited a farm at Frome, Ontario, and she and Dad moved back to the St. Thomas area to look after the property. Dad began working on the Michigan Central Railroad as a locomotive fireman and later as a locomotive engineer. They lived in a home on Chestnut Street.

In 1918, Dad was appointed full-time general chairman of the Brotherhood of Locomotive Firemen and Engineers Union. This was a different organization from the railroad and resulted in Dad being gainfully employed during the terrible Depression years. We were able to move to a larger home on a three-acre fenced piece of property in Yarmouth Heights, a suburb of St. Thomas, Ontario.

In 1948 Dad was elected vice president for Canada of the Canadian Brotherhood of Locomotive Firemen and Enginemen and National Legislative Representative in Ottawa, Ontario. Mother and Dad moved to Ottawa and retired there in 1959.

During their marriage, Mother and Dad always traveled together except for the brief periods when Mother was home giving birth to their children. They were very close and the majority of their time was spent together. Mother and Dad both loved to hunt and fish. They owned a lodge on an island in Asagami Lake in the Sudbury area where they spent their summers. Mother passed away April 27, 1970, and we were not surprised that Dad soon followed on June 30, 1970.

Looking back, we were an undisciplined group of kids while our parents were away. They always left us in capable hands, but it seemed that every time they returned home they had to get one or all of us out of any messes we had gotten ourselves into. They always complained about how tired they were of seeing and having to listen to the sheriff when they got home.

Growing Up

In Yarmouth Heights, where we lived, we didn't have a police force, and the only law enforcement was the county sheriff, Mr. Lefty Prevail. His job was to keep the peace in our area, and he always tried to settle the different disputes and problems in a reasonable way. He would interpret the law and make judgments that would be agreeable to all parties concerned. In fact, the sheriff was well liked and respected by everyone. There were a few incidents where I managed to trigger a visit from the sheriff.

I had a large paper route delivering the *St. Thomas Times Journal* to the whole area where we lived. A boy my age by the name of Roy Trigger had a vicious dog. Every time I delivered the paper to his house, he always had his dog out and running loose. It would always try to bite my pants and paper. One day, the dog was totally uncontrollable, and it ripped my pants and bit me several times on the leg. Roy didn't say anything during all this, so I told him I would be back in about fifteen minutes to fix him and his dog.

Enough was enough! I went straight home and untied Bum, my German shepherd dog that I had raised since he was a pup. I hooked his leash onto the handlebars on my bike and went back to Roy's house. The dog was outside, and Roy and his mother were sitting on the front porch. I told Bum to sic him, and Bum went after the other dog. They got into a terrible fight. Roy and I also got to fighting, and I knocked out his front tooth. While all this was going on, Roy's mother was screaming and hollering, saying she was going to call the sheriff. Hearing this, I grabbed my bike and Bum and headed home.

While my mother put iodine on my dog bites, I did my own hollering and screaming. Into this melee came the sheriff, Lefty Prevail, and Roy's dad. Mr. Trigger told my dad and the sheriff that their dog had to be put down because of his injuries, that Roy had to have a bridge made to replace his missing front tooth, and that my dad would have to pay for it. We then gave the sheriff our side of the story, and the sheriff told Mr. Trigger that they shouldn't have had such a vicious dog running loose and that the McLean's didn't have to pay for his son's lost tooth. He said the matter was settled as long as I didn't get an infection in my legs. Nothing further came of this incident.

Beside us was the Dayton Davis farm. They had an old cannery on their property with a lot of windows in it, and they used the building as a barn. One day, my brother Robert and I both thought it would be fun to get out the .22 rifle and shoot out some of the windows. As we were shooting the rifle, we realized how quiet it was, so I suggested we get out the shotgun and give it a try. Mr. Davis heard the shots and called the sheriff.

When Mr. Davis and the sheriff turned up at our home, I said, "Robert, I told you we shouldn't have gotten out the shotgun."

Robert was so upset by what I said that he was sputtering, and I finally owned up to the fact that it was me who suggested we get out the shotgun.

Since I had a job, I agreed to pay for the windows, because I didn't want my parents to find out what we had done. I settled everything up with Mr. Davis.

On the Davis farm they raised thousands of chickens as well as sheep. It seemed that every night, some of the chickens got through the fence into our yard. I would tell Bum to get the chickens, and he would grab them by the neck and shake them till they were dead. He got to go back several times a day for nibbles. This didn't seem to me to be a problem, because the chickens were on our property. I didn't realize the danger of having a dog that killed.

One night Bum got into our barn and caused a terrible commotion. Mother woke me up to take care of the problem. Bum had killed all twelve of my Banty roosters that I had raised. The noise was coming from my pet crow screaming in the rafters and Bum barking, trying to catch him. I still didn't realize the danger Bum presented.

One day I was on our tennis court trying to build a propeller-driven snowmobile when Bum jumped on my back and knocked me into a steel pipe I was working on. I got a large cut on my arm, and Bum kept trying to lick the blood off my arm. I had to keep pushing him away from me so I could get into the house. Even at this time I didn't realize how dangerous he was becoming.

I had heard rumors that sheep in the area were being killed, but I didn't associate the killings with my dog. Several mornings when I went outside, Bum had broken his chain and been free all night. Then, one day, I found pieces of wool around his mouth, and I knew he was the one killing the sheep. I didn't tell anyone and thought that if I just kept him well chained to his doghouse, he wouldn't be able to escape from our yard.

One morning I found Bum was missing, and Mr. Davis was in our yard talking to my Dad. Earlier that morning, Mr. Davis had caught Bum in his chicken coop killing some of his chickens and had shot him. He told us that eleven of his sheep were dead and that Bum was the culprit. He also said he had called the sheriff and that Dad would have to pay for the loss of his sheep and chickens.

When the sheriff finally arrived to settle the situation, he checked his agriculture book, which stated that if Mr. Davis killed the dog, my parents didn't have to pay restitution, because the by-law clearly stated that he had to hold the dog until it was proven that he was the one who had actually killed his sheep.

The story of Bum ended there. It was finally brought home to me that Bum had become a very dangerous animal that had to be put down and that he should have been put down a lot sooner. It was scary to think what might have happened if Bum had turned on me or someone else.

There is one more incident I remember and want to mention.

When I started in secondary school, I was on the boxing team for Arthur Voaden Vocational School. I was just a mediocre boxer, and a boy by the name of Ed Compton was the champion during this time. When I was growing up, the only thing my Dad ever asked me to do was to make sure I won the match that was coming up between Ed and me.

Ed's father did nothing but brag about his son and how good he was. I guess my Dad was getting tired of hearing this, and he wanted me to win for his sake.

"You're always fighting anyway, Peter," he said, "so you should be able to win."

I took it all to heart and talked to George O'Neal, the boxing coach. He told me the most important thing in boxing was to have strong legs and to be very light on your feet. I knew if I was going to win the boxing match, I would have to work hard and do some serious training.

I took to skipping rope, practicing enough to be able to skip every which way, and got so that I was very good at it. Ed Compton was taller than I was, and he would win his bouts with his long reach. Nobody could get close enough to hit him really hard.

When the boxing match came up and we started to box, I just kept my head down and kept pounding him in the stomach. Because of my newly learned fancy footwork, all the blows he gave me just grazed the top of my head, and I kept hammering him in his stomach for the first two rounds. Then, in the third round, his stomach was hurting him so bad that he was bent over, and it gave me a chance to really get some good blows to his head. One punch knocked him down, and he couldn't get up for the count. I had won the match!

When I got home that night, Dad already knew the outcome of the fight.

I asked him how he knew what had happened. Without my knowledge, he had been at the match and watched it from the back corner.

He said, "You did a terrific job, Peter. Now I won't have to listen to Mr. Compton brag about his son anymore."

He now had his own bragging rights.

Sibling Tales

Bill (William) was the oldest in the family, and he was Mother's favorite. She would always make sure that he had Dad's car to use at night, and she would supply him with money for gas or whatever. In Mother's eyes, Bill could do no wrong.

Bill and his friends found out how to make hard cider, and Bill got a forty-gallon wooden barrel that he put in the basement. He would fill it with cider and create a mixture in the barrel that would turn the cider alcoholic. He had his own personal still. Bill and his friends would tap into the barrel whenever they went out. Dorothy, our oldest sister, tried to stop it, but to no avail. Mother would put her foot down and tell everyone not to criticize Bill and to leave him alone.

An aside to Bill's hard cider—being a large family, we needed lots of water. One time Dad hired a well digger named Eaton to dig us another well. Eaton loved the hard cider Bill made and knew where he kept in the cellar. While he was digging the well, he kept sneaking into the cellar to have a few drinks. No one at the house oversaw him or even thought to see how much he accomplished in one day or how much he was drinking. There was one day, however, when he had too much to drink and thought that he was an expert piano player, and he decided to play on the piano in our living room. He climbed out of the hole he was digging and went into the house, sat down, and started to play Mother's piano.

When Mother saw Eaton in his dirty clothes playing on her piano, she started screaming and hollering at him to get out, and she chased him right off the property. When Dad came home, however, he had to rehire Eaton to finish digging the well. Mother was not pleased but reluctantly agreed to his coming back as long as there was no drinking and he would be watched from then on.

Years later there was one episode with Bill that I'll always remember. Bill was visiting Mother, and he was sick with pneumonia, and started to get the DTs from his excessive drinking. He kept thinking there was a radio station on top of the house and he could hear voices and music. He was so perplexed that Mother decided she had to call the family doctor to look at him and tell her how to handle him.

Dr. Potts came about two in the morning and stayed for about two hours. Nothing he did helped, and finally he decided that Bill needed to be admitted to the nearby psychiatric hospital to dry out. Bill refused to go anywhere unless I agreed to go with him. Mother called me to come to the house, and she packed clean clothes for Bill, along with a new pair of pajamas. I'll never forget those pajamas.

When we arrived at the hospital, the charge nurse said she was expecting us and to just follow her to our room. She took us to a semi-dark dormitory where there was a small light over two single beds. The nurse left us there without giving us any instructions.

I told Bill that since we were there for the night, we might as well get into bed. He refused to get into bed unless I did, and he wanted to know which bed I preferred so he could get in the adjoining bed. Bill took off all his clothes and put on his new pajamas. He looked neat as a pin, and when he got into bed, he looked just as sane and in as good of shape as I was. I whipped off my shirt but left my pants and shoes on, pulled down the covers, and jumped into bed. Bill immediately fell asleep and started snoring.

While lying there, I began to get nervous, thinking the nurse wouldn't realize which one of us needed the help, and I thought I was in one hell of a fix. Here I was, half-dressed, and Bill was in his new pajamas, looking quite peaceful. I couldn't sleep. I kept thinking the nurse would keep me and send Bill home, but she returned at seven in the morning and said I could go home and that Bill would be just fine. I jumped into my clothes and left in a hurry.

Bill was in hospital for about three months. After two months, I went to see him and thought he was getting squared away mentally, so when he called me to pick him up after three months, I thought everything was okay with him.

Bill and his first wife, Velma, had three daughters, Charlotte, Mary, and Elizabeth, as well as a son who had died very young. At that time, Bill and Velma were separated, so Bill had nowhere to go. The first place he wanted to go to was the Royal Hotel to have a few beers. I tried to check him into a room at the hotel, but he said not to bother because he was staying right where he was in the bar.

We had been there about two hours when a woman came in who was a nurse in London. She knew Bill had no place to live or stay and that he was bound and bent on drinking again. The woman decided to take Bill home with her so she could take care of him. He lived with her for about a year.

Bill was an expert auto mechanic and could always get a job. He was at a Ford dealership in London and got himself back on his feet financially. However, he never stopped drinking.

On one of my visits, I arrived at the dealership about noon, figuring he would be ready for lunch. He said he was glad to see me and that the job was getting on his nerves. There was a farmer standing nearby who Bill thought was watching his every move, and he was tired of it. He hung up his tools and locked his box.

We went to the Brunswick Hotel, and he decided he was not going back to work that day. When the dealership finally got a hold of him the next day, he told them he quit right then and there.

Bill was the type of man who every woman wanted to take care of, and he soon took up with a woman named Olive from Simcoe, Ontario, who owned nursing homes in Simcoe and Niagara Falls. They lived in the nursing home in Simcoe, and when his divorce from Velma was finalized, he married Olive.

Bill used to go to the bars in Niagara Falls, New York, for nights of partying. One night when he was there, as he was crossing the street on a green light, some fellow ran the red and hit him, breaking both of his hips. It was a terrible accident, and he was confined to the hospital for three months.

One day Olive called me to come to Niagara Falls to help straighten Bill out. Apparently, friends had been bringing him liquor, and he had been doing so much drinking that the hospital decided they could no longer take care of him. By the time I got there, Bill had been sent home, and Olive had decided to put him in her nursing home in Niagara Falls, Ontario, until his hips healed. It took some time, but eventually they healed, and he ended up walking with just a slight limp.

While he was recuperating, Bill wrote his papers to become a stationary engineer, which he passed easily since at one time he had worked on steam engines on the railroad when he was working as a fireman. He and Olive were separated by this time.

By then, Mother and Dad had moved to Ottawa, and Bill went there to live with them for a while until he got squared around again.

Soon after, Bill got a job taking care of a large apartment complex, which required a stationary engineer to look after the boilers and refrigeration, because the building was steam heated. He had this job for about five years and was given his own apartment in addition to his salary.

He was divorced from Olive by this time and he never again married.

He finally got a settlement from his accident in Niagara Falls. I can't remember exactly how much it was, but he got quite a substantial amount of money, and it enabled him to live and maintain himself in excellent style.

When he was about fifty-five years old, Bill died peacefully in his sleep.

Dorothy (Jane) was the oldest girl and never married. She always said it was because of growing up with five brothers she had to help raise with our parents gone so much.

Dorothy was given the task of helping to supervise the rest of us. She and Bill fought constantly. Bill would not listen to anything she had to say. Dorothy had red hair and the temper that went with it. One night at the dinner table, she got so mad at Bill that she threw the ketchup bottle at him. The bottle missed and broke when it hit the wall, creating an awful mess. The room not only had to be cleaned but also totally repainted, and it had to be finished before our parents got home.

The cider in the basement was another bone of contention between Dorothy and Bill. She was always emphatic that the hard cider had to go, but Bill would say, "No way," and Mother would always stand up for Bill.

Dorothy finished university and ended up in real estate in Ottawa where professionally she was called Jane. She was president of the Ottawa Real Estate Board for many years and did very well financially. She was well liked and respected by everyone who knew her.

June was the third sibling, and at an early age, she married Tommy Johnstone, and they had four boys, Robert, Tucker, Johnny, and Jerry.

June always worked hard in whatever business they were involved in. Eventually, they built a lovely restaurant on the property where their original icehouse had been. They sold their home and moved into an apartment they built over the restaurant. June did most of the cooking, and Tommy ran everything else. Everyone thought the business was thriving. Life seemed good for them. Their four boys were grown up and had all left home.

One day Tommy just up and left. He had sold everything, the business and the building they were in. No one thought Tommy would do such a thing, but there was my sister, June, left with nothing. All we knew was that Tommy had taken everything and moved to Florida.

I thought June would fall to pieces, but to my complete surprise, she didn't. I was very proud and admired her for how strong she was and how she handled things. She found herself an apartment, got a job at an auto dealership, and made a life for herself.

Eventually she divorced Tommy and then remarried. Her new husband, Bill Plastow, was known for being so tight with his money that it was like bark to a tree, and he liked to gossip. Not many people liked him, but he was good to June, and I guess that was all that mattered.

June was doing well, and at the same time, Tommy had been diagnosed with cancer and was in a hospital in Florida. They shipped him home to a hospital in St. Thomas, where he died alone.

After Bill Plastow died, June moved into a seniors' residence, but after a couple of years, she got sick and had to be hospitalized. She died shortly thereafter.

Robert was two years younger than me, and because of the shortage of bedrooms in our house, I had to sleep with him in a double bed. This is what started our problems. If Robert so much as touched me, I would go crazy, and a fight would begin. One day Dad got so fed up with us fighting that he picked us both up and threw us over the back porch railing into a snow bank. We stopped fighting quickly.

Any time we were having a "social gathering" at Mother's, she would always tell me ahead of time that under no circumstances was I to get into a fight with Robert.

Mother was having a party to celebrate Robert's graduation from the Air Force at Haggersville as a pilot, and she again told me emphatically that under no circumstances was I to get into a fight with Robert.

I don't remember who touched whom first, but a fight started. It ended with Robert's uniform being all torn up. Robert's wife, Marg, hit me on the head with one of her high-heeled shoes, and I got a cut on my forehead that bled all over us before the fighting stopped and we made peace. I apologized to my mother and explained it wasn't my fault and that I had tried to avoid it, but Robert wouldn't leave me alone. At least that was my take on this particular fight.

When Robert was drinking, a fight between us would usually start. I've never really understood the problem, but it was there our entire life.

To further illustrate that the feelings between Robert and me continued throughout the years, I cite the following from when we were in our forties.

One time I had my Mother, Dad, and Robert, along with one of his friends, up to my property on Batiste Lake during hunting season. Ted Boomer, my best friend and sometime business partner, was with me. We were all enjoying ourselves, and the weather was good. Ted came and told me that Robert was filling his liquor bottle out of the liquor supply I kept in my room and this was the reason my supply was shrinking.

After a few days, Robert and his friend said they had to get back home and that Robert's car was about fifteen miles away from where we were. I told Robert that Ted and I would drive them out in our four-wheel-drive Land Rover. I told Ted that I had to watch out for Robert, because I knew what would happen if he got to drinking before we left. Ted didn't believe me.

We were traveling on a bush road heading for Robert's car, and I was sitting in the backseat with Robert's friend. Ted was driving, and Robert was in the passenger seat. We were talking and laughing, and all of a sudden Robert turned around and grabbed me by the shirt collar and started shaking me and yelling, "I could take you right here and now if I wanted to."

Ted got to laughing so hard he couldn't steer the truck, and we ended up in the ditch and got stuck in the mud.

I got out to check for damage and told Robert to grab the axe in the back of the truck and to cut down a small tree to use under the front tires. Robert went ahead and did what I told him to do, and he forgot all about fighting. We all had to work hard to get the truck back on the road before we could get Robert to his car. Later, Ted told me he never would have believed this could happen if I hadn't told him about it beforehand.

In our senior years, Robert and I socialized more, and when not drinking, we got along well.

Robert graduated from the University of Western Ontario with a minor in geology, but he went into the insurance business and did very well. He never did do anything connected to mining.

Robert and his wife, Marg, had a son, Mike, and a daughter, Peg.

There was a large age gap between my two younger brothers, Gordon and Russell (whom Mother called Mike), and me. Since I ran away at such an early age and married young, I was not around the house very much while they were growing up, so I hadn't interacted with them like I did with Bill, Dorothy, Robert, and June.

Top: Family Home on Fairview Ave in 1933
Bottom: Mother & Dad's 50th Wedding Anniversary 1961
L-R: Myself, Mike, June, Dad, Mother, Bill, Dorothy, Robert, Gordon

GETTING STARTED

New York Central Railroad

With my early fascination with trains, it was fitting that my Dad got me a job as a machinist apprentice on the New York Central Railroad. It was October 1, 1937 and I was nineteen years old.

When I first started, my shift began at 7:00 a.m., and every morning I had trouble getting out of bed to get to work on time. I used alarm clocks and even tried setting an alarm clock in a dishpan so it would create a lot of noise when it went off. That seemed to be the biggest problem that I had with my job. A nineteen year old looking for a good time doesn't consider the hours he keeps.

Finally, my Dad told me that if I didn't want the job enough to get to work on time, then he wanted me to quit. After a couple of months or so, I solved the problem by getting to bed earlier. This-early-to-bed-early-to-rise routine became a habit I have kept until this day. What a novel idea!

The apprenticeship took four years to complete, and I made journeyman machinist on June 30, 1941, and then became an assistant house foreman the next day.

During the war years, my position with the railroad was classified as "number-one priority for essential services" for the war effort. Because of this, I was unable to enlist in the military, although I did try several times.

We had a conscripted labor force on the railroad during the war years. I had lawyers, farmers, barbers, housewives, beauticians—all different types of workers—in my engine house. They were needed to help maintain the steam engines, and they all had to be trained and guided.

During these years, it was vital that the trains run on time, and it was my job to see that they did.

I felt I had the world by the tail with all the money and time to do what I wanted. I had no one to be accountable to except myself. I was still young, and plans for my future were not fully developed. I knew, however, that every step in life was the start of a new adventure.

Certificate of Apprenticeship, Journeyman Machinist 1941

Marion and Me

In the 1930s and 1940s, Port Stanley, just south of St. Thomas, was the place to be in the summertime. It was the largest and most popular resort on the North Shore of Lake Erie and people came from far and wide. There was a special electric train that ran from London to St. Thomas to Port Stanley daily, and it was always at full capacity or had standing room only. They also ran an incline railroad to service the people who lived and stayed in the hills around the town.

Port Stanley had a boardwalk with all the popular amusements and a wonderful beach area. On the boardwalk was everyone's favorite place called Mackie's, where they sold the greatest hot dogs, hamburgers, fries and their special orange drink.

It also boasted about getting all the big bands to play at the world famous Stork Club, and they drew large crowds. They had a large dance floor that was mounted on springs to make it soft and flexible for dancing, and with a lot of people on it, the floor would sway. The pavilion held anywhere from two to three thousand people, and we danced to Guy Lombardo, Glen Miller, and many, many others.

Like others my age, I was out for a good time, and a good time I had, with no restrictions. Eventually I got a girl pregnant, and my first thought was for her to get an abortion. There was a doctor in the area that performed abortions although it was illegal, and I brought the girl to him to be taken care of. Then, there was a second girl pregnant. When I brought the second girl to the doctor, he wouldn't do the procedure unless I stayed right there with him and watched the whole thing. It was worse than anything I could have ever imagined. He said that this was the last abortion he would ever do for me.

In early 1938, I met my wife, Marion, at the Stork Club and took a complete shine to her. Marion was originally going there because of another fellow who played in the band. I think his name was Doug Baker. I discovered that Marion had a wonderful singing voice and could have done quite well if she had done it professionally. But I decided that Marion was the one I wanted to marry, and I pursued her relentlessly. We would dance a lot at the club and spent most of the year going out together.

In January 1939, we found out that Marion was pregnant. I then had a major decision to make. Would I look for another doctor, or would I just up and leave like some men I knew did? But I was in love with her, so we went to Port Stanley and got married in a private ceremony on January 14, 1939. I had just turned twenty-one years old.

We hid the marriage from our parents for a short time. We needed to figure out where we were going to live. Then, one day, Marion's Dad found our marriage certificate in a drawer where Marion had hid it, and he went to talk with my parents about our situation. Between them, they found a suitable place for Marion and me to live.

Finally, when I talked to my Mother and Dad about being married, my Mother said it was okay if that was what I wanted. She also told me, "If you want to have good luck and good fortune, you have to look after your wife and family."

I never forgot her advice and tried to practice it in my own very erratic manner throughout the years. I do have to admit, however, that there were affairs, but I never thought about ending my marriage.

Our life started out quite good because I had just inherited $500 and was working as an apprentice on the New York Central Railroad. Marion worked at the Monarch Knitting Company, and we did well financially for a short time. I'll never forget, though, that Marion needed dental work, and it cost us the $500 I had inherited.

Author Peter James McLean   Marion Attwood McLean
Pictures taken in 1938
I always thought I looked like a gangster in these clothes
Marion was a blue-eyed beauty

Toy Guns

My mind has never been settled on one thing or on one job. Even at a young age, I worked on making a snowmobile, but it never held my attention, because all I wanted to do at that time was run away from home and become a sailor. However, when I finally settled down and got married, I was always thinking up new ideas and inventions. I couldn't wait to get started on each new adventure.

It was during this time on the railroad that I invented and patented an assortment of toy guns. I built different models of automatic elastic guns and automatic cork guns that were made from aluminum and plastic. I had a good assortment of models made up. Later, an opportunity would come up where I could follow through on these inventions.

Toy Guns Invented and Patented in 1947
Hand Made Bazooka that shot cord projectiles
Prototype Automatic Elastic Gun

Model Steam Engines

Also, while learning my trade, I had all the machinery and resources I needed in the shop that I could use after hours, and I began to build

Peter James McLean

things. By hand, I built a range of model steam engines—twelve in total—that could run with compressed air. Today, the steam engines have been totally reconditioned by Gordon S. Tuck who is one of the few people in Ontario that has a steam boilers license for steam engines and locomotives.

The steam engines are presently in climate controlled storage until such time as the railroad museum completes their heated show room in St. Thomas.

Some of the Model Steam Engines I built
Keeping them working for Grandchildren

Discovering New Uses For Magnets

I continued thinking up new ideas and inventions, and my first real success was in using magnets. During my early studies and research, I came across a short article about magnets being used in Russia on stationary boilers to prevent lime scale and hard water buildup. I found it very interesting because I was thinking of all the trouble we had with this problem on our steam engine boilers. The boilers had to be washed out once a month, and it was a big job. You had to remove all the washout plugs on the boiler and wash the inside with water treatment chemicals to remove the lime scale, hard water, and impurities in the water from the boilers. This represented at least eight hours of work and labor to clean each one of them.

After reading the article, I was thinking it would be a good idea for magnets to be tried and tested on at least one of the steam engine boilers on the railroad. On my own, I began experimenting to figure out how to make up a set of magnets for one steam engine boiler and see if I could get it working. I found a magnet manufacturer in Toronto and had them make up a set of magnets for me. The first dozen sets were not large enough and not the correct configuration to fit on the water intake pipes of a steam engine boiler.

I finally developed a set of large magnets made up in a *V* form and in two portions so that they could be clamped on the pipes. They had to be made so that the north pole and the south pole were opposite on the magnets when they were installed in proper alignment.

I put the new magnets on one of the steam engines I was servicing and ran it for a month. Before the boiler workers started washing the boilers, I used to blow off cock (one on each side of the boiler), and I blew the water down just by holding the blow off cocks. The boilermakers removed the plugs and inspected the inside of the boiler for lime scale, hard water, and mud buildup. It turned out that the mud ring of the boiler was clean with no buildup. In fact, the boiler was the cleanest that the boilermakers had ever seen.

I showed the master mechanic how the magnets would solve all the labor problems of washing out the boilers on the engines manually by just using the blow off cocks to remove all the sludge and the hard water build up in the boiler. I said it would save the water treatment and acids that had to be used for the type of boiler washout that we were now doing on a monthly basis. He said that even though it looked good and sounded good, it was necessary to have the federal government inspector come and examine it and verify it worked as shown before we could start using them on all the steam engines that we were involved in servicing.

Federal Inspector

The federal inspector for the railroad in St. Thomas was a man by the name of Wilf Wiseman, and he and I got to be good friends through the years. Wilf would usually come into the shop to do his regular inspections every two or three months.

When Wilf came for his next inspection, he examined the magnets and the procedures I was using. He said, "Peter, it's tremendous, and I'm going to okay it."

I then had more magnets made up, and we used them on about a dozen steam engines that were assigned to the Canadian division. These were the engines that the Canadian maintenance serviced, and we did the monthly inspections on them. We used the magnets on the locomotives, and it saved approximately eight hours of labor for each boiler.

In those days, if you created a money-saving procedure for a company, they didn't reward you with money or a bonus. However, they did give me a "certificate of merit" in honor of the project.

Wilf had bought a farm in Havelock, Ontario, and every time he came to the shop for his inspections, he would always need something for the farm. Every trip it would be something different, and it was an unwritten understanding that when Wilf was booking federal defects on the locomotives that I was involved in, he would overlook the majority of defects that he could have booked. This made my job much easier. The federal inspector's word was law. If he looked at a boxcar or locomotive, it was within his power to book any defects that he noticed—and when he booked something, it was the law of the land that it had to be taken care of and repaired immediately. No train he inspected left the yard without his approval.

On one trip, Wilf came in and said, "Pete, my car's not running worth a damn, and I'd like you to put it in the shop and have one of your mechanics go at it and try to find and repair the problem."

I said, "Okay, Wilf, we will if it's not too serious, but we'll find out what's wrong with it."

The mechanic found out the carburetor was plugged and needed to be cleaned and that he would have to order some new parts for it because we didn't have the parts needed on hand.

The master mechanic was a man named Cheney who had just been assigned to St. Thomas from the United States, and he didn't realize how important the federal inspector was or even what he did. Cheney saw the car in the pit with two mechanics working on it and said, "Okay, Pete, get that car out of here; this is not an auto repair shop. I don't want any more work done on it."

The most important train that we ran through our division was the hotshot freight train referred to as the JS2 It was loaded with perishables and other goods bound for New York City. The train was such an important freight train that it had to run on time just like a passenger train.

It was necessary for me to take a crew down the yard and dump the hoppers on the locomotive engine, clean out the front end, and check that the engine was running good. It was so important that we didn't cut the engine off; we did the work on the engines right in the yard on the fly.

As soon as the work was suspended on Wilf's car, there was suddenly a red flag put on the JS2 because Wilf Wiseman had found several defects on the boxcars of the train. The red flags meant that the train couldn't move until the defects were taken care of.

Immediately, the phones started to ring, and everybody wanted to know what had happened to JS2. The chief dispatcher said we were in one hell of a jam. Brass from Detroit was on the phone, and the head brass out of New York was also calling.

Wiseman told the superintendent, "JS2 doesn't move until these defects are taken care of. You got my car held up, and it is as important as the JS2."

The superintendent got me on the blower and said, "McLean, McLean, fix Wiseman's car. What does it need?"

I told him, "The carburetor has to be rebuilt, and some parts have to be ordered."

He told me to do it immediately, and he then got a hold of the inspector and said, "Wilf, I didn't realize the importance of your car. McLean has got two or three mechanics working on it right now, and they are repairing it as fast as they can. We have to work something out with JS2 to get it moving, or we'll all be fired."

Wiseman said, "Okay, some of these defects on the boxcars can be transferred to Harmon, New York, and I will send the work orders there to make sure the work is taken care of."

So the JS2 was released out of St. Thomas and continued toward New York. I had Wilf's car fixed that day.

After that episode, Cheney, the master mechanic, nearly got fired.

Nothing like that ever happened again. The superintendent and master mechanic in St. Thomas learned quickly that when dealing with the federal inspector, his word was law. Part of their job, therefore, was keeping the inspector happy.

Certificate of Merit 1948

Interesting New Hire

One day, Dude Lock, master mechanic on the CNO Railroad, came over and asked me if I wanted to hire a good machinist.

"Sure, we could certainly use one."

"Well, let me just tell you about him. His name is Steve Stedman, and right now he is doing ninety days in jail for stealing approximately $10,000 worth of material from our stores department. But Steve is a very good mechanic, and he is a willing worker."

"Well, he's just a complete thief, eh, Dude?"

"Yes, but outside of that he's very trustworthy. I like him very much, and with winter and Christmas coming, he has a family to take care of, and I'd like to find him a job."

"Does he use a gun or anything?"

"Steve is completely nonviolent, and I'd really appreciate it if you would hire him."

"Sure, I'll take a chance. Send him over when he gets out of jail in another week or so."

Steve showed up on the day we agreed on, and I had a good talk with him and told him there would be no thefts while on our railroad property here in St. Thomas.

Steve was very likeable, and I put him to work doing one of the dirtiest jobs we had. He was to pull flues out of the boiler on an engine that was in for repair. Steve went to work with two helpers, and he did the job very well. He turned out to be a very conscientious worker and would willingly do any job I gave him.

I always left my cigarettes and lighter on my office desk, and one day my shiny metallic lighter went missing. I went to Steve and told him my lighter was missing and that I couldn't light my cigarettes. Steve was the only one I told about my missing lighter, and when it appeared on my desk within a half an hour, I knew that Steve was the one who took it in the first place. I never mentioned it again.

About a week later, another worker, Bill Bear, came to tell me that his jar of silver change that he kept in his locker for coffee, etc., was missing. I again told Steve there couldn't be any thefts in the round house. Shortly after, Bill came to say he found his change jar sitting on his bench. I realized then that Steve was a real kleptomaniac and couldn't resist taking anything that was shiny.

Another time, I remember Harold Askew, who worked on the gas cars that we serviced, came to me one day to say that the $200 worth of socket wrenches he had just bought had disappeared from his locker. I told Harold to give me some time and that I'd find out what happened to his tools. As usual, I went directly to Steve and told him that if the wrenches weren't found, I would have to call the police to investigate. A couple of hours later, Harold came to say the wrenches were on the floor under his workbench.

Regardless of all this, Steve was such a good and willing worker that I couldn't stay mad at him.

One day, Steve showed up at work with a brand-new saxophone along with all the accessories, which must have cost over $1,000. I asked him, "What are you doing with it?"

Steve said, "I always wished I could play the saxophone in a band."

He then asked if I would talk to Clinker Clinesmith, the commander of the legion band in St. Thomas, about him and his saxophone.

The next time Clinesmith came in, I introduced him to Steve and had Steve show him his new saxophone.

Steve said to him, "I would love to play in a band. Do you need a sax player?"

Clinker said, "Yes, indeed, I do need a sax player, and I would be glad to have you join us. In fact, the band will be playing this Friday night at the legion, but you would need a uniform."

Steve told him, "That won't be a problem. I'll be there!"

Clinker and his band were part of a competition that included bands from Chatham, St. Thomas, and the London area, and that Friday night was the final for the competition. He wanted his band to win and thought Steve and his sax could help.

The next time I saw Steve I asked him, "How did it go with the band?"

He replied, "Everything went just great, and we won the competition."

At noon that same day, in came a very agitated Clinker who wanted to see Steve Stedman right away. The man who had sat beside Steve in the band came to Clinker and told him, "The joker sitting beside me couldn't play a note of music."

Clinker wanted to find out for himself if Stedman could or could not play the saxophone. Clinker took Steve into the lunchroom and asked him to play certain notes on the sax, and it didn't take long for Clinker to realize that Steve couldn't play anything.

Clinker was beside himself. He said his band had just won first place in the competition, and here he had a member who couldn't play a note. He didn't know how to handle the situation. I told him that he should just leave things alone and accept his first place prize.

Clinker never forgot winning that competition with a sax player who couldn't play a note!

After this episode, the stealing became less and less frequent—it was just shiny stuff that Steve couldn't resist.

Steve kept his job until he was laid off with many of the other railroaders. The last I heard of him, he had a job in the parts department of an automotive dealership in St. Thomas.

One of the Steam Engines we serviced in St. Thomas

City Council

In the early 1940's, I decided I wanted to try my hand at politics, so I ran for alderman on the St. Thomas City Council and won the election. I served one term and was appointed chairman of the health unit. Because of the war and my job on the railroad being classified as "essential services," I couldn't take the time required to do the job properly, and I never again ran for city council. This small taste of politics, however, soured me from seeking any other elected political office.

Railroad Union

I worked as a supervisor in different areas until a new union contract was put into effect in 1948. Under this new contract, supervisors only had to work one day a month to keep their seniority and pension benefits, and I took full advantage of this in order to work on outside interests.

My frequent absences were possible because I always had an associate anxious to fill in for me on the job. I would arrange with him that I was going to be gone for a couple of weeks, and he would fill in for me. They would always say yes, because they wanted my higher rate of pay.

Transfer to Windsor

I was not having any luck in St. Thomas attracting investors for my different inventions, and it wasn't long before I was running low on funds and needed to do something.

In 1948, when I learned I was being transferred to Windsor as the new engine house supervisor, I jumped at the chance to move. I was looking forward to the change in environment where I could meet new people who could help me financially.

However, there was one major glitch. Marion did not want to move to Windsor. We ended up separating for a time. We had two daughters then—Barbara, born in 1939, and Carolyn, born in 1942. I went ahead anyway and made the move by myself.

THE EARLY YEARS
IN WINDSOR

Railroad

The general superintendent in Windsor for the New York Central Railroad was a man by the name of George Bunclark. He had a very bad reputation as being a strict disciplinarian when it came to rules and regulations. All my railroad friends in St. Thomas said I would not be happy working for him.

When I arrived in Windsor and met George, I was assigned to the midnight shift, from midnight to 8:00 a.m. From the very beginning, I never had a disagreement with George, and he never criticized in any way the job I was doing in the engine house. He was extremely pleased with the production of the midnight shift with its reduced staff.

I would like to explain the pressure we were always under working at the Windsor Engine House. No steam engine could go through the railroad tunnel from Windsor to Detroit, Michigan, and vice versa. Every steam engine arriving in Windsor from the Suspension Bridge (in Buffalo, New York) had to be serviced. That meant the engine had to be unhooked and brought into the engine house for repair. Once the steam engine was unhooked, the train would be hooked up with an electric engine for the trip through the tunnel.

The boys at the Suspension Bridge had the habit of sending all the poor engines to Windsor with the idea of letting the Canadians do all the repair work required. It seemed that every steam engine that came into Windsor was not steaming and was running very badly. The work sheets the engine crews made out always reflected this.

To do this work properly, you had to dump the fire and send two boilermakers into the firebox. They would have to clean the flue sheet by hand, because they were always covered and plugged with scale. This was the main cause of the engine not steaming. This required three or four hours of manual labor in order to do the job properly.

The shortcut I developed was to use the deck hose, sometimes without even dumping the fire. The high-powered force of the ice-cold water was squirted in the firebox and onto the flue sheets. It would snap and crackle, and the scale would explode right off. After doing this for about fifteen minutes, you would have a clean flue on the steam engine so that it would produce steam perfectly and be ready for its eastbound run.

This operation was not legal or permitted according to federal rules and regulations because it was hard on the flue sheets and could possibly start them leaking. If they did leak, it would be very little, and there was never any danger to the boiler itself.

There was one other shortcut that I used on these low-steam engines that came in from the Suspension Bridge. I would clean the front end and put passenger coal in the steam engine, and with the clean flue sheets, it was ready to go. I was able to service a steam engine and have it ready for its return trip within about one hour.

From the Windsor yards to Tilbury it was an uphill grade. After the engine left our yard, I would phone the dispatcher later and ask him how JS2, WB4, or 358 passenger trains were running and what speed they were going when they passed Tilbury. The dispatcher would say they passed Tilbury at sixty miles per hour, and you would know that the engine was running good and the eastbound trip would be very successful.

George Bunclark never asked me any particulars on how I was able to repair the steam engines in order to have them running so good on the return trip. If anything was ever mentioned about it, I would just say the front end was plugged and the engine had a poor load of freight coal on it.

Down at the coal chutes, we had one chute for freight coal, which was very poor coal, and then we had one for passenger coal so the engine would run cleaner. I had my men using passenger coal on the engines.

This high-grade coal helped the engine get up a full head of steam as it went through the countryside, and the dispatchers would love it because the trains were running at the proper speed limit in order to make good time.

I found George Bunclark to be the most honest and upright person that you could ever work for. The reports I had gotten about him before coming to Windsor turned out to be untrue. My working for him as an engine house supervisor was always very successful, and I never experienced any problems with management.

Personal Life

My private life in Windsor without Marion and the girls was not very stable or successful.

I had rented a house on Hall Avenue and filled it with furnishings from Tepperman's Furniture Store with a small down payment and small monthly payments to carry it. The house on Hall Avenue didn't work out, and the furniture was repossessed by Tepperman's. I still had a balance owing that I had to make payments on.

It was while I was still living in this house on Hall Avenue that I had brought my oldest daughter, Barbara, to Windsor for a visit. She was about nine years old. My brother Bill was in town and asked me to drive him back to Ottawa, and I thought it was a good idea. But Barbara was still with me. So what did we do? Well, we just brought Barb with us on our trip.

The only problem was that we kept stopping at bars along the way for a few drinks. Although we left Barb in the car while we were in the bars, she didn't appear to be upset, because to make her happy we gave her anything she wanted. In our minds, this scenario worked just fine. By the time we finally got to Mother's in Ottawa, Bill and I were so intoxicated we could barely stand up. Barb, on the other hand, was tired but in good spirits because she was about to see her favorite grandmother.

This was the only time I can remember Mother being totally outraged at Bill (and me). She couldn't believe we had done something so stupid as to drive drunk with Barbara in the car. Or that we would also leave her alone in the car while we went into the various bars along the way for a few drinks. Mother made certain that by the time we sobered up we fully understood that what we had done was not only outrageous but dangerous as well.

I never forgot Mother's reaction to our drunk driving. From then on, I could say I never drove drunk with one of my children in the car. But when you are intoxicated, there are things you don't remember. So I can only hope I never again drove drunk and endangered one of my children or other drivers. But as you will read, I did drive drunk again and again.

On my way back to Windsor, I dropped Barb off in St. Thomas and then returned to Windsor.

It was at this point that I moved in with a Swedish couple on Janette Avenue, and I spent my time drinking and working. I started drinking right after work in the morning and didn't get to bed until really late in the afternoon, and then I had to get up and go to work again at midnight. It was a very hard life, because I was not sleeping or eating properly. In fact, I was self-destructing on my own in Windsor.

Besides owing money on the furniture, I had bought a brand-new 1948 Chevy and had it financed through Household Finance in Windsor. I finally made up my mind that things were not going to work out without Marion and the girls with me, so I was determined to patch up my marriage.

It was not easy getting Marion to make the move from St. Thomas, where she was living with her parents, and I had to convince both her and her parents that I could keep the family together in Windsor. Carolyn was staying with her mother, and Barbara was living with family friends. Barbara hated the people she was living with, and Carolyn kept running away from school. They both needed stability. Marion finally consented and moved with me to Windsor. I finally had my family back together, but my money troubles were not over.

I put an ad in the Windsor Star to sell the Chevy, and it sold immediately, but the payments on the car were in arrears at Household Finance. About a week after selling the car, there was a knock on the door, and a man said that he had come about the car. I thought he wanted to buy the car, and I told him it was already sold. He then said he was from the finance company and he was there about the payments I still owed on the car. He thought I had broken the law by selling the car, but I told him that it wasn't illegal since I still intended to keep up the payments. Over and above what I got when I sold the car, there was still some money owing on it.

Now I had payments to make on the car I had just sold along with the payments owed on the furniture that was repossessed. With my wages and the payments I had to make, I barely had enough money left for me to be able to feed my family and pay the rent on our apartment.

My job on the railroad was going well, but I decided to drive a cab for extra money and went to work for Veteran's Cab Company. I would drive the taxi after I got home from work, and I soon realized how important the tips were. I certainly needed the extra money.

One day I picked up five American tourists at the Detroit/Windsor tunnel, and they had me drive them to the Airport Hotel on Walker Road, which turned out to be a bawdy house. As I was talking to the madam, she said that for every gentleman I brought to the Airport Hotel, she would pay me five dollars. So when I was driving the taxi on Saturday or Sunday nights, I would tell the male passengers I picked up at the tunnel that I knew of a great bawdy house, and I asked if they wanted to go there. The men never refused. After about a year, from the money earned driving the taxi together with the tips I received, I was finally out of debt.

Certificates and Diplomas

In the 1930s and early '40s, quitting school early was very acceptable, because a lot of kids had to work to help finance the family. At the time I ran away from home at sixteen, I didn't give a thought to more education. I thought I knew everything necessary to do what I wanted in life. However, as I was doing my apprenticeship to be a journeyman machinist on the railroad, I came to realize that education was indeed important.

At that time, I could not afford to quit working to attend school full time, and there were no night classes offered. I took certified courses through the Railroad and International Correspondence School that prepared me for promotion at work and enabled me to pursue other interests over the years. All the diplomas and certificates I received also made me look better on paper.

The following is a listing of the various certificates and diplomas I earned over the years:

Railroad

    1941: Certificate of apprenticeship, journeyman machinist
    1945: Assistant engine house foreman certificate of merit

International Correspondence School

    1943: Diploma, locomotive machinist (twenty-five courses)
           Diploma, locomotive brake equipment (five courses)
           Diploma, business correspondence (six courses)
    1945: Certificate in mathematics (ten courses)
    1947: Diploma, foremanship (twelve courses)

Dale Carnegie Institute

    1946: Certificate in effective speaking, personality development, and human relations

Other

    1948: Dominion of Canada patent for my automatic toy guns
    1953: Certificate from the State of Michigan Corporation and Securities Commission's Guy Underwater Exploration Club
   *1955: Certificate of ordination from the Gospel Ministry, member and license (five degrees)
    1968: Lifetime prospector's license, the Mining Act
    1969: Graduated from the Electro-Motive Division of General Motors at LaGrange, Illinois, and became a diesel instructor for Penn Central Railroad

1980: Granted honorary title of admiral of the Great Navy of the State of Nebraska from then-Governor Charles Thone

1981: Certificate of membership, Society of American Inventors

1984: The Toronto Futures Exchange Board of Governors; successful completion of the TFE futures floor traders examination and offer of assistance in the process of purchasing or leasing a seat on the exchange

*My thinking at the time was that I could become an evangelical minister. I even thought about the sermons I would give. I decided that if I limited myself to preaching about the Ten Commandments only, then I wouldn't have to read the whole Bible. Obviously, I quickly lost interest.

Remember the old saying "You'll never know everything about everything, but there will always be someone who knows everything about one thing"? Always go to the experts and ask for help. This is another rule I lived by.

I cannot say enough about how important education is in your field of interest. You can never stop learning.

Research

After finally getting my family settled in Windsor, I had easy access to the Detroit Public Library. I spent many, many hours there doing research, and the staff couldn't have been more helpful. In addition, they had a special telephone line called "T.I.P.S." that you could call and find out any information you wanted on anything. It was the perfect resource. It was like having your own personal research assistant.

This, of course, was before computers.

The 1957 movie *Desk Set* starring Spencer Tracy and Katharine Hepburn told the story of a research assistant in a large company who was trying to keep her job when the company wanted to replace her by bringing in computers. This was similar to using the T.I.P.S. hotline at that time.

To this day, I have never stopped learning and continue to do a tremendous amount of research. There is always something new out there to learn, and my grandson in later years fixed me up with my own computer to help me compile information. Although I missed having a real person to talk to and look up information for me, the computer was a tremendous help, and I slowly started using it more and more.

Ted Boomer

There I was, living in Windsor, and though my finances were very low, I was finally out of debt. I had managed to find us a place to live in the upper apartment of a duplex in the two hundred block of Janette Avenue. As Marion and I got settled in Windsor, I kept thinking about the various ideas I had running around in my head and tried to settle on a new adventure to begin. Soon fate intervened when another family moved into the ground floor unit.

The new people who moved in below us were Ted and Helen Boomer. Ted and I introduced ourselves and started talking about what we were each working on. I told him about the toy guns I had invented and patented, and he was very interested in them after I showed him the prototypes.

Ted was born in Hamilton, Ontario, in 1917. He moved to Windsor before entering the war in 1939. He served during WWII as a sergeant in the Tank Corps.

After the war, Ted returned to Windsor and founded a new taxi company, Veterans Cab Company. It was the first cab company to have radio dispatching, and before long, other cab companies followed his lead.

Ted had sold the cab business to drill for oil on Pelee Island. At the time I was working for Veterans Cab, Ted had already sold the business.

Ted then related to me what had recently happened to him. He said he had just gone broke on Pelee Island and that it was his own fault. Ted continued with his tale:

"I was drilling for oil on the island and had two financiers and their lawyer flying in from Cleveland, Ohio. I told my wife that this was going to be a very important meeting because the men were looking to buy out the whole operation. I had lots of liquor in the cottage so they could have anything they wanted to drink. I filled an empty Crown Royal bottle with cold tea and told Helen that when I asked for a drink she was to give me the tea. However, the guests were to be served whatever they wanted.

"The meeting started about noon, and we were discussing the drilling and the progress and the amount of money that I needed for them to buy me out lock, stock, and barrel. The meeting went on until seven or eight that evening, and the highest price that they offered was $200,000. The men were quite intoxicated, but since I was only drinking tea, I was sober. I thought I could manipulate them into accepting my asking price of $250,000. They insisted that they could not go over the $200,000 they had already offered me.

"I kept thinking that when I hit oil it would be worth $2 to $3 million, so I gave them a flat refusal, and the meeting broke up and the men flew back to Cleveland.

"I was still drilling over the next three and a half days. All of a sudden, at about 3:00 p.m., the well blew in, and it turned out it was natural gas, not oil. Natural gas on Pelee Island was not worth a nickel. I had invested all my personal money into the project, and now I have nothing left."

Ted and I quickly became friends for life, and he was part of many of my adventures over the years.

The two of us managed to get into a lot of trouble together because of our drinking.

INVENTIONS BROUGHT TO LIFE

Planet Police Toy Guns

Ted Boomer was the best salesman I had ever met. At the time we met, we were both broke. Beer, however, only cost ten cents, so we would scrounge up two dollars to have a few drinks.

From the moment I showed Ted the guns and balloons I had invented, he said they were the greatest things he had ever seen. It was during our drinking sessions that we hatched a new adventure. He thought we should form a company and put these toys on the market. We figured we could pay ourselves a salary out the investment money we could collect.

We investigated the toy market and found there were no automatic projectile guns on the market. These were the first ones ever made for children to play with. So, in my spare time, Ted and I formed a company.

Ted knew a lady, Mrs. Ida Clark, who lived here in Windsor and had money. Ted thought she might be interested in investing in our project. He took me to meet her to show her what we were doing. She liked the toy guns and invested enough seed money for us to get the company started so we could sell shares and raise capital.

The guns turned out to be very expensive to get to the manufacturing stage. To get a metal die made to manufacture one complete gun at a time cost around $25,000 (1948–1949 prices) for a plastic injection mold die. Then we had the expense of making the packaging and display units. We had to raise more money through the selling of shares. The company was very popular, and the people we talked to loved the guns and the program we had set up to get them on the market.

"Planet Police" was the name we decided to use because of the growing popularity of science fiction and aliens, and so it became the trademark name on the patent. The company was named Peter McLean Enterprises. We had the packaging, helmets, guns, projectiles, and display units with pictures on them. We had the guns shooting small plastic projectiles so it was harmless for children and they could use it for target practice. I also had a gun that would shoot elastics.

When we had our first prototypes ready, we decided to take them to the International Toy Show in New York City. We had a nice setup and display of our products. In fact, they were so popular that we won first prize for the most innovative new toy on the market. This was the beginning of the automatic toy guns that shoot multiple projectiles.

While at the toy show, orders came in from buyers by the thousands, and Ted and I both knew then that we had a huge problem. We had all these orders and couldn't fill them. Our injection die mold was just making one gun at a time, and we had orders for thousands. It would take us a year to get new dies and the manufacturing process set up to where we could make at least two dozen guns at a time.

In reality, we should have had the manufacturing process set up and a warehouse full of the guns ready to ship before we even went to the toy show. The other toy companies went right to work and brought out their own version of our toy guns shortly after. Our company could not compete.

We had the patents and copyrights, but we didn't have the finances to proceed with any type of lawsuit against the large companies.

After meeting Ted and getting the toy company started, I felt the hard times were over, even though this company was not successful.

Marion and I moved again into the upper apartment in a duplex on Dougall Avenue. One day, Marion and I were sitting in the living room watching our very first television, and one of my automatic elastic toy guns was being advertised. Marion said that I needed to sue those people, because they were using my design.

I had to explain to her that suing would cost us money we didn't have and that we couldn't win against the large toy companies.

Peter James McLean

This was my first and last foray into the toy market.

Right: Picture of myself holding toy gun prototype
Left: Planet Police Display at New York Toy Show
Bottom: Planet Police Packaging designed for Toy Show

Additions to the Family

It was during this time that my oldest son, James Guy, was born in 1951, and our youngest son, Peter Jesse, was born in 1952. I kept telling Marion that if the boys were born in December, I could use them as a deduction on my income taxes. Both boys were born in January.

Marion wanted me to buy a home because we were now had four children. When the girls were small we moved around frequently. My daughter, Carolyn, often said that by the time she was in Grade nine, she had attended nine different schools between St. Thomas and Windsor.

On my own, I purchased a home near the corner of Peter Street and Rosedale Blvd. Marion was happy with the house and we settled in. The house had a garage but I wanted it bigger so I built onto it and made it as large as our City Bi-Laws permitted. This became my new den.

Marion was a stay-at-home mother. She was well known for her baking, and excellent cooking skills. At one time she baked cakes, pies, and cookies and sold them to family and friends. She had no trouble selling everything she made.

Also, she once took a cake-decorating course and won first prize in a competition. She made a heart-shaped two-layer cake and called it "Hearts and Flowers." It was beautiful. Our whole family remembers the decorated cakes she made to celebrate all our birthdays and any other occasion that came up. Marion was especially good at doing wedding cakes, and she did a lot of them for friends and relatives. In fact, she made a decorated cake for our daughter Carolyn's wedding. We always seemed to have company for dinner on Sundays, and no one ever refused her special desserts.

Overall, I found Windsor to be a very lucky town for me, and I never again felt that I would suffer financially. It wasn't long before I began another new adventure.

Las Vegas and Reno

In 1957, Marion and I decided to take a trip through the United States from Detroit, Michigan, to California. I decided to drive, and we took Route 66. We felt it would be a nice way to explore the western US states. I wanted to see Dodge City and the other outlaw locations.

We stopped and explored everywhere and everything we found interesting, and then we arrived in Las Vegas. We spent two days there, and I tried all the gambling machines when not sightseeing. However, I didn't win anything, just spent a lot of money.

After exploring what we wanted to see in California, we arrived in Reno. We stayed in a motel, and we got up early the next morning and decided to go to Harrah's Club, the biggest casino in town. We got to Harrah's about eight in the morning, and the place was wide-open and booming.

I began to play the slot machines and right off the bat I got two or three jackpots. Then I started to play keno and card games and again started winning. Everything I tried I ended up winning. I had a jacket on with big pockets that were full of winnings along with buckets that were full.

Somehow the casino found out my name, and one of the hostesses was following me around, calling me "Mr. McLean." How she ever found out my name I'll never know. She kept following me around, saying, "Mr. McLean, what would you like to drink?"

I kept telling her, "Club and water would be fine with me."

So everywhere I went, she was there with a tray holding Canadian Club and water for me to drink. I didn't know that the casinos in those days supplied free drinks to people who were winning.

This had never happened to me before. I was having tremendous luck, but with Marion's coaching we stopped gambling, because I was getting too intoxicated to drive. We went back to our motel, and when we counted the money I had won, we found that everything—the entire trip—was paid for with quite a bit of money left over.

We then drove to the Nevada border and stopped at a bus depot there. Marion and I went in for lunch and a couple of drinks for me to help settle my nerves. The people were so wonderful and so friendly, and we were having such a good time that I thought it would be a good idea to share my winnings.

I ended up giving everyone in the bus depot fifty dollars to try to bring them some of my luck. Nobody could believe that I was giving away this money, and I told them that I just had to share my winnings because nobody should be that lucky. There was still a lot of money left over. People just have to understand that I didn't consider this charity. As I said, it was my way of sharing in the hope that my good luck would continue.

The next day we arrived in Steamboat Springs, and the snow was so deep you had to have extension poles on your car's antennae so you could be seen above the snowdrifts. From there to Denver, the drive was mostly uphill. It was about a two-hour drive. I tried to buy some chains for the Cadillac's tires, but they were all sold out. A gas station operator told me that there were a lot of people struck in the mountains. We were traveling Highway 40, a main highway, and the road was partially sanded but not very good.

I told Marion we were going to go ahead and make the drive without chains. When we started out, the back wheels were spinning, and the speedometer said eighty miles per hour, but really we were just going about thirty-five or forty. We were very lucky. There was no traffic, and I never had to stop or slow down along the way. We finally cleared the top of the mountains and started down toward Denver.

After about ten minutes on the downward trip, a fellow in an Oldsmobile was on the wrong side of the road, and I had to swerve over into a large snow bank in order to miss hitting him. He stopped and thanked me for not hitting his car and he said, "I've got a shovel to help you dig the Cadillac out of the snow bank."

As I was shoveling the snow and clearing my car, I was panting and puffing to beat the band, and I said to the fellow, "I must be in bad shape. My condition must be very poor."

"No," he said, "what's going on is the altitude. We are so high up that the air is thin, and we aren't getting the oxygen that we should be getting." Here I was thinking I was having a heart attack.

We got the car out of the snow bank, and being on a downward grade, we had an easy trip down the mountain and got into Denver safely.

There we were, sitting on the outdoor patio of a restaurant in Denver. The sun was shining and it was warm as toast. I told Marion I couldn't believe how we came from blizzard conditions into the sunshine.

Road trip to Las Vegas and Reno

Dr. Richard Dubinsky

Rick was in elementary school with my two sons and he enjoyed spending time in my workshop while I was doing all my experimenting. I may have been drinking heavily during those years, but I was always glad to see Rick.

One experiment that he was really excited about at this early age was growing gems using an apparatus that I had bought from the National Research Council of Canada. I would show him how the machine grew sapphires and different types of gems depending on the formula used. At that time, making synthetic gems was very mysterious.

Rick was very helpful when I was experimenting with the Gas-O-Rator, trying to figure out how to get higher production out of the oxygen and hydrogen to supply more volume.

Rick was interested in anything having to do with science, and he eventually graduated with a PhD in physics. Over the years, he worked with various companies, as well as at NASA and in other government positions.

Rick was always interested in what I was doing, but I could not get him interested in prospecting. He came up to my property with me a few times, but he said he never seemed to have time for prospecting. He thought that prospecting was much harder than it first appeared, and he never got gold fever.

To this day, Rick comes for a visit whenever he can. He enjoys talking about my inventions and other projects that could be done in the future.

Gas-O-Rator

Around 1962, while doing research, I got interested in water. It was then that I realized that water was one of the most mysterious substances on earth.

At the time, while I was growing sapphires and gem material, I had been using hydrogen and oxygen in the burners. Then I began to realize the explosive power in a gallon of water in comparison to a gallon of gasoline. The water had a thousand times more explosive power!

I also began to experiment with the electrolysis method of breaking the water down into hydrogen and oxygen. The improvement I made was to use nickel screening with the positive and negative extrudes that I required in breaking down the hydrogen and oxygen from the water. I named the apparatus the "Gas-O-Rator," and I used a high-mileage 1955 Cadillac for testing. It was a vehicle I owned but was not using at that time, and it was perfect for what I needed.

I kept improving the Gas-O-Rator on the Cadillac until I was getting approximately fifty miles per gallon of gas.

I began to talk to people about my new invention and the improvements that I had made in gas mileage and the operation of the car motor. I found out that feeding the oxygen and hydrogen mixture from the Gas-O-Rator into the motor left the inside of the motor very clean, and when you examined the spark plugs, they would look just like new, with no carbon deposits on them.

The hydrogen and oxygen was the perfect combination for the explosive power inside the cylinder along with the gasoline. It would give you additional power, and it improved the gas mileage.

It was a tremendous idea, and it worked very well on my car. When the oxygen and hydrogen exploded in the cylinder, it would turn into small amounts of steam and then evaporate out through the exhaust pipes.

I thought that I was involved with a million-dollar-plus invention and the Gas-O-Rator should sell extremely well to the public. Everybody, it seemed, was looking to improve his or her gas mileage.

However, I soon discovered that the government, automotive companies, and the oil industry were not at all interested in my invention.

I contacted the head office of the Ford Motor Company in Dearborn, Michigan, and they asked that I bring my car over and have an interview with the vice president of engineering. On the day of the appointment, I drove my 1955 Cadillac with the Gas-O-Rator installed on it to Dearborn. I had put a padlock on the hood that I had mounted so that no one could open the hood without my permission.

When I got up to the vice president's office, his secretary said I had to sign a document before my discussion with the VP. I read the papers, which in effect said that after the meeting I would have no protection on the ideas that I was presenting to him. I told them I couldn't sign the papers without discussing them with Ted Griffith, my patent attorney in downtown Detroit.

I got my lawyer on the telephone and read him the papers they wanted me to sign. He told me that if I signed the papers, all my rights and my ownership of the invention would be null and void and that Ford Motor Company, if it wished, could proceed with their own development of the Gas-O-Rator. Needless to say, I did not sign the papers, and so our meeting was canceled.

While I was upstairs at Ford Motor Company, someone tried to pry open the hood of my Cadillac, and they bent a portion of the hood where it was padlocked. However, they were unable to break the lock.

I then contacted Chrysler Corporation, and representatives said they were very interested. They said they were going to send two of their engineers to my residence in Windsor so they could test the device on their own automobile.

Ted Boomer and I had an appointment at my place at approximately two in the afternoon on a certain day. The two engineers came at the appointed time. However, they were driving a big Chrysler car that was in terrible shape. The engine was missing on two or three cylinders. The exhaust pipes were smoking, and the car was burning oil.

I commented that the men certainly brought an awful car to test my invention. They said they were thinking that if the Gas-O-Rator worked on their car, it would work on any car, so they were here to give it a test.

I mounted the equipment on their car, and we took it for a test run. Highway 401 from Windsor to Tilbury was finished at the time, and we felt that distance would give us a great test run.

I told Ted to drive and that during the drive I wanted him to floor the car and have it wide open as much as possible. Ted got behind the wheel, and I sat in the backseat. One of the engineers sat beside Ted, and the other one sat in the back with me. Ted couldn't resist telling them again that the car they brought was a piece of shit.

Ted filled up the car at the gas station, and we headed out to Tilbury. As soon as we got on the highway, Ted opened the car up, and our top speed was about seventy miles per hour. That was as fast as the car would go, and there were sparks coming out of the exhaust from the carbon being blown out of the engine. We got it up to eighty miles per hour. I told Ted to keep it wide-open, and we got it to ninety miles per hour. Before we got to Tilbury, the car was running at a top speed of 120 miles per hour.

We filled the car with gas in Tilbury, and Ted drove back home the same way. When we got home, we filled the car with gas again and then drove into my driveway. We were using the gas tank to determine mileage.

On the drive to Tilbury, the mileage worked out to between eighteen and twenty miles per gallon. On the return trip, however, the mileage worked out at fifty-one miles per gallon. The two engineers commented on how smooth the car was running, and they came across as being very excited about the Gas-O-Rator. They said they would be reporting to their superiors and that we were going to hear back from them shortly.

Ted and I both thought the two engineers were pretty rough-looking characters. I told Ted that as a precaution I was going to put everything in the basement of my house and would let him know as soon as I heard from Chrysler. I was waiting to hear what kind of a deal they would make.

The next night, my garage and den were broken into and ransacked. I called Ted and had him come over to see the damage that had been done.

I had steel bars on the garage windows, and someone had taken one of the steel bars off a back window and gotten in that way. Nothing had been stolen. The only thing Ted and I could figure out as we were talking was that the boys from Chrysler had come back and broken into the garage to steal the Gas-O-Rator. I was relieved that I had thought to put the equipment in the house.

During the following week, Garvey Shearon, my friend and the assistant manager of the Bank of Montreal in downtown Windsor, called and told me that there had been inquiries about my bank accounts and my banking business. He said it was very suspicious to him and wanted to know what was going on, so I told him what had transpired.

A short time later, Ted and I were sitting in the Norton Palmer Hotel, having a couple of beers and discussing the Gas-O-Rator and what was going on. We had the oil companies and the government both against us, and we had never heard anything back from Chrysler, so they certainly weren't going to offer us a deal. We decided it was time to put a halt to this adventure.

Oil companies did not want more fuel-efficient cars, because it meant less gasoline sold and less tax revenue for the government. The car companies would do nothing to upset either one.

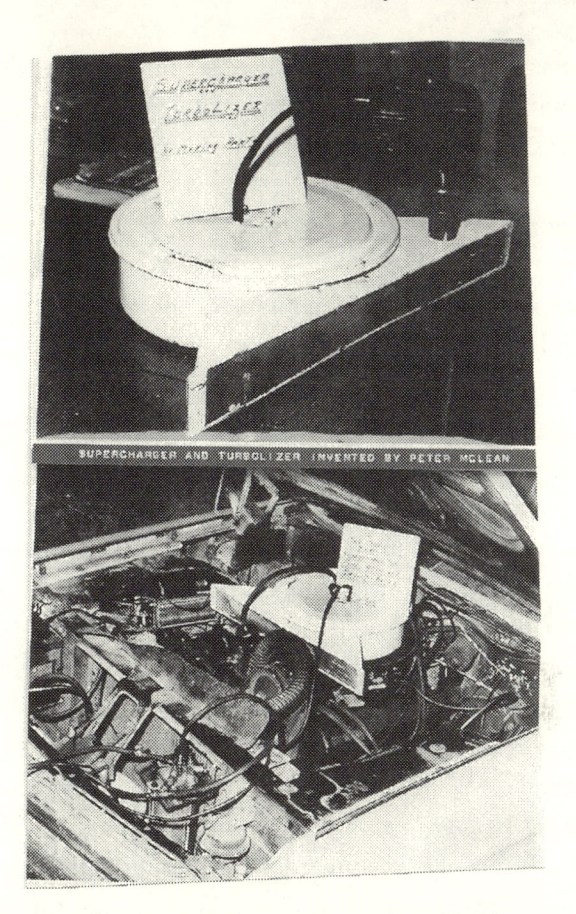

Running Gas-O-Rator on one of my Cadillacs

Oil Purifier and Filtered Cigarette Holder

While developing my Gas-O-Rator, I was also working on an oil purifier that would maintain viscosity. Instead of having to change your oil every five thousand miles, this oil purifier would keep the engine in such good condition that you would not have to change your oil for well over one hundred thousand miles.

In May 1965, I received a patent for my oil purifier unit, but after the problems with the Gas-O-Rator, I didn't feel that it would be given any more acceptance, so I just filed it away.

I also invented and patented a cigarette holder that filtered smoke through water during this time period, but again, I didn't think it was an exciting invention, so I never followed up on it.

Electro-Magnetic Water Purifier

After having success with the magnets on the steam engine boilers at the railroad, I began to think that these magnets would be perfect for household use because of the benefits. It would provide softer water so soap and shampoo would lather better as well as prevent the scaling of kettles and showerheads. It would provide better drinking water, all at no operating costs.

I had my Toronto manufacturer make up smaller *V*-shaped magnetic models to be used on household piping. I formed a company called Electro Magnetic Inc. and made up folders and packaging with money out of my own pocket. I didn't have any shareholders in this company, so the start-up was very expensive to me personally. The people of Essex, Ontario, had hard water, and I sold units in that area. They were selling very well. Also, the water in Essex had a sulfur smell to it, and my product did help with this problem.

I spent of a lot of my own time and money selling the magnets. The people who had the units said they worked well for softening the water for their bathing and laundry because they were now doing these things in soft water.

The one detriment to the magnets is that after snapping them on the intake water pipes, you just left them there, and they worked 24-7 with no mechanical moving parts. Because there were no moving parts, the magnets just didn't have the appeal to the public that I thought they would.

People equate noise with power—like the gas engine. They have to hear the noise and/or see moving parts to be interested. Even to me, this didn't prove to be a very exciting project.

It was time again for me to find a new adventure.

THE BOY'S CLUB

Marion did not approve of my drinking, so I could not keep liquor in the house. That left my garage/den, and I drank there alone many times. When Marion was on the warpath about my drinking, she would pour any liquor she found down the drain. What a waste! To hide my liquor, I would sometimes stash my bottles in the bushes of my close neighbors.

Can you imagine in this day and age an inebriated man roaming the neighborhood, sometimes in his boxer shorts, looking for his liquor stash? Our neighborhood consisted of many professional people, but somehow I got away with it at the time.

I never brought my drinking friends home. I tried very hard to keep my home life and drinking life separate.

When not drinking in my garage, I spent my time drinking at my favorite club. It was at Ma Ellis' house (a blind pig), which was located downtown next to the police station. The people who frequented this establishment, referred to it as "The Club".

The following were some of the interesting members I drank with.

George Yates

My close neighbor in Windsor was a lawyer named George Yates whom I had first met at our club. Shortly after meeting him George got convicted of drunk driving and lost his driver's license for one year because of an automobile accident that he had caused while driving drunk. In those days they didn't have Breathalyzers so when you were convicted it was never for impaired driving. It was always called drunk driving.

Whenever I was available and it didn't interfere with my activities, George would ask me as a favor to drive him different places.

George was a very good lawyer with a large practice and it turned out that we became good friends and associated with each other all during the years I was in Windsor. He became my personal lawyer and served me well at the various times he was needed.

It was a lot of fun and enjoyable for me for the different times I would take George to the Courthouse when he had a case on the go. In the Courthouse in Windsor they have a lawyer's room for them to change their clothes and keep their files, use the telephone and whatever.

It always made me laugh that the most important package in the lawyer's lounge was a bottle of liquor, and it seemed that the majority of the lawyers would always have a strong drink in order for them to be squared up and be able to talk properly in the courtroom. It seemed that George would always need a stiff drink or two to help with his hangover from the night before. It helped to calm his nerves and clear his speech so that he was presentable before the Judge.

A woman by the name of Ruth had been George's secretary for many years, but she left George's employment and had moved to Sault Ste. Marie, Ontario to look after her mother, who was in a nursing home. While there, Ruth got a job with the city handling some finances, and she had taken some of the money and used it for her own personal use. Ruth was charged with theft. She contacted George and asked him if he could come up to the Sault and represent her.

George called me to ask when I was making my next trip up north to Wawa. Since I had to pass through the Sault to get there, he asked if I could take him and drop him off there. I told him I was going the following Sunday and would be glad to take him with me. He said he would be bringing Marion Clark, his girlfriend, with him and that they would be returning to Windsor by train.

The weather was clear when we left Windsor even though it was wintertime and we had good driving conditions. However, when we got in the vicinity of Clare, Michigan, where there was a curve and bridge on the expressway, the car got loose, and we ended up in a big snow bank in the ditch. It turned out that when this happened, we were one of about four cars that were also in the ditch.

A tow truck was already working on the other cars. The tow truck operator told us to get in the truck and that he'd drop us off at a tavern within a quarter mile of where we were. He said he would bring the car to us when he got it pulled out and then we would be all set to go.

When we got started again, George said his nerves were so bad that we had to find a couple of bottles of liquor to take with us, which we did. He and Marion were drinking all the way to the motel in the Sault.

The next morning, George called my room about eight thirty and asked, "Peter, will you please do me a favor and call the district attorney's office and tell him that I'm unavoidably delayed and will be in the courthouse at 10:30 a.m.?"

This was not a court case that was being held; it was just a meeting George was to have with the prosecuting attorney about Ruth and the facts regarding the case.

George was very hung over because of all the drinking the day and night before, and he wanted to hit a liquor store at ten in the morning and grab a pint of Canadian Club that he could keep in his pocket. He said he needed a few drinks on his way to the meeting.

George and I went to the liquor store, got the liquor, and then went to the courthouse. We stopped in the men's room to have a few shots before the meeting. George wanted me to come in with him to kind of break up the meeting a little bit and give him a chance to get started off right and maybe arrange some sympathy for Ruth. George's goal was to get the prosecutor to go easy on Ruth, and he thought he would be able to soften him up.

Into the office we went, and before George could say anything, the prosecutor started off by saying, "George, you look very familiar to me."

George laughed and said, "It's possible since we're both lawyers."

When they asked each other what university they had gone to, they each said, "Queens." They had gone to the same law school.

Then the prosecutor told George that they had been in the same class at university. George said he was starting to remember. They each asked if the other remembered this or that and began talking about all their activities and the fun they had during their college days. Apparently George and the prosecutor had been really the best of friends at the time. This back-and-forth went one for quite some time, and finally, the talk got around to Ruth.

The prosecutor told George, "Ruth has asked to repay the money. One way we can settle this is for me to give Ruth her job back and make deductions from her paycheck in order to repay the money in full. It's not going to be necessary to have any kind of court hearing over this case."

George said, "I really appreciate your doing that for Ruth, because she has fallen on hard times and was having family problems with her sick mother in the nursing home." It turned out there was very little discussion over Ruth or her problems when the two lawyers finished talking.

That was how quickly they settled the case. It was that simple.

By this time, George really needed a drink and started making some hints saying so.

"Now, look what I have here," said the prosecutor as he opened his desk drawer and pulled out a forty-ounce bottle of Canadian Club. "We're all going to drink on the success of this meeting."

We spent well over an hour in his office with George and him having a good time drinking and talking about old times. When we got back to the Hotel, George said, "My god, Peter, that was the easiest case I have ever handled and Ruth is all taken care of now."

This is just an illustration of how lawyers and the wheels of justice can work to the advantage of people in the know.

I drove on to Wawa, and George and Marion took the train back to Windsor.

George had met Marion in the Killarny Tavern, where she was the hostess. They finally moved in together in George's apartment in a building that he owned on University Avenue. Shortly after he got his divorce from his first wife, he married Marion Clark. They got along very well over the years, because George could drink in front of her and she would accompany him to different outings where drinking was involved. From the first time George introduced me to Marion, I couldn't get over how much she looked like his first wife.

George's first wife was very critical of his drinking. There were numerous times when George would ask me to bring him a bottle of liquor, discretely hidden in a newspaper, up to his bedroom in order for him to sneak a few drinks to help him get squared up while he was getting dressed in the morning. He lived only two houses from me. George's house was the coldest house I had ever been in. There was never anything out of place, and it was very unwelcoming. It was almost like nobody had ever lived there.

One afternoon when George and I were sitting in the Norton Palmer Hotel Bar downtown, a fellow came to our table, and George introduced him to me. He said he was representing this man in court, but I didn't have to leave unless I wanted to. When he told me the gentleman's name, I recognized it immediately, because his name had been in the *Windsor Star* for at least a week.

The article in the newspaper was about him being the biggest dope and narcotics dealer in Windsor. He had just gotten out of Kingston Penitentiary a short time before and was being charged again for dealing dope.

When I realized who the man was, it jarred me, and I realized that I didn't want to be associated in any way with such a character. I told George I had something else to do and that I had to leave. We said good-bye and I left.

Later, I told George I didn't want to be near those types of clients. He said he had no problem because he was the man's lawyer. But for me, I didn't even want to be seen at the same table as them, and that's why I skedaddled right out of there. I kept getting myself into enough trouble without associating with known criminals.

At one time, the management at the Norton Palmer for some reason thought I was a ball player. They were always giving me complimentary rooms.

There was a lovely lady who worked there as a hostess, and she and I began an affair. It didn't last very long, but her husband found out about it, and he gave the woman a real beating. In fact, he ended up breaking both of her legs.

There were no police involved in this incident, and we never heard anything about her or her marriage again.

George was an excellent lawyer, and he handled most of my legal problems over the years.

Norm Riordon

Another club member that is worth mentioning was Norm Riordon. He was a mild-mannered lawyer who had a very nice personality. He had an excellent record of winning cases for his clients. His specialty was citing the law. He used the various loopholes and technicalities that he would find in court to show the reason his clients were not guilty.

Everybody liked Norm because of his obliging personality. I knew of a lot of cases that he had handled where the clients did not have the money to pay him in full for his services. But Norm always said that the people he represented would be able to pay his fees at a later time.

Norm always felt that a client going to court was under a lot of pressure, and he sympathized with them and felt they should have good legal representation. During this time, there was no legal aid, so if the lawyer wasn't paid, it would be money out of his or her own pocket.

Norm had an office on the seventh floor of the Canada Building in downtown Windsor, Ontario. His practice seemed to always be busy and profitable for him. He was a widower, having lost his wife several years earlier, and he had no outside obligations.

Needing a new secretary, Norm hired one by the name of Diane. She was tall, blonde, and very nice looking and was approximately fifteen years younger than he was. Once Diane became Norm's secretary, she began to show her true colors. She soon began to take complete control of Norm, his office, and his law practice.

After getting to know her, you realized that she had a heart of stone and that she was determined to marry a lawyer to improve her social standing. It didn't take her long to win Norm's heart. They got married, and it was the worst thing that could ever have happened to poor Norman.

Norm had a lovely home on Rosedale Boulevard that was three houses around the corner from where I lived and right next door to George Yates. Our backyards overlooked each other.

Confessions of an Eccentric Dreamer

Diane moved into Norm's home, and soon after, she got Norm to close his business office in the Canada Building and open an office in the basement of their home. Her aim was to cut down on the expense of maintaining a downtown office.

Soon after, we began to hear screaming and hollering and noisy behavior coming from Norm's home. The arguments and quarrels started shortly after they were married.

Norm's practice quickly began to go downhill after moving his office. He was no longer handy and available for his clients.

Every evening, Norm would spend his time at the Dominion House, a bar on Sandwich Street about three blocks from his home. If you went there in the evenings around ten or eleven at night, Norm would be sitting there drinking beer.

During one confidential discussion with Norm in the Dominion House, he said, "My marriage to Diane was a terrible mistake, and my home life is becoming hell on earth. The only way out of this terrible mess is to divorce Diane as soon as possible. I need to get my life straightened out and try to get it back to what it was."

Soon after, Norm and I were in the club one morning, and Norm said, "Peter, would you please go and move my car off the street and hide it back in the alley?"

"Sure I will, but why do you want me to do that?"

"I'm so depressed I don't even want to go out into the street in the daytime."

Finally, Norm filed the papers to divorce Diane, and he ended up moving out of his home into an apartment downtown on Pitt Street. It appeared that one of the provisions of the divorce proceedings was that Diane got the home on Rosedale.

Norm was even more despondent than we realized, and, shortly after he committed suicide. He was such a neat and careful person that even when he committed suicide he climbed in the bathtub at his apartment and wrapped himself in towels before he shot himself so he wouldn't leave a large mess to be cleaned up.

One of the high spots of Norm's life that he always liked to talk about was his good friend Gene Autry, and he would often show us a picture of the two of them. Norm had met him in his younger days when he lived out west in Calgary, Alberta. Norm remained friends with Gene throughout the years, and whenever there was a parade or celebration held in the Windsor/Detroit area with Gene Autry as the main guest, Norm was always invited to ride with him in his vehicle during the parade or attend the event.

About a month after his death, his home on Rosedale was sold, and Diane seemed to disappear. Nobody had heard anything about her until a year later. George Yates got a phone call one day saying Diane was in Grace Hospital on University Avenue and that she wanted George to come and see her.

George called me and said, "I hate that bitch, and I want you, Pete, to drive me over the hospital. We'll go up together to see Diane and get the story of what's going on."

When we got there, we found out Diane had taken an overdose of some drugs, presumably to kill herself. She now wanted George to represent her in a case about a person who had taken advantage of her. This person had spent and drank up all her money, and at the present time she was destitute.

George said, "Diane, you need to explain everything so I can understand the full story behind this episode."

Diane said that she had a boyfriend who had been living with her in the LeGoyeau, which was at that time a deluxe apartment building down on the Riverfront at the corner of Goyeau Street and Riverside Drive. She said that she financed all their activities, including lending the boyfriend substantial amounts of money during the relationship.

Diane told us that the previous Saturday morning there was a knock on their door, and when she answered it, there was a lady standing there with two children.

Diane asked the woman, "What can I do for you? What do you want?"

"I've come to get my husband back."

"There is no husband staying here that I know about."

The boyfriend got up out of bed when he heard the commotion at the door. He went and spoke to the woman and then told Diane that he had to leave. He got dressed and left immediately with his wife and children.

George asked, "Who is this guy?"

Diane said, "He is an electrician from Detroit who told me that he wasn't married. It turns out though that the woman who had come to my door is, in fact, his legal wife, and they were his children who were with her."

During their entire association and relationship, the man had sworn up and down to her that he wasn't married and that he was a single man with a good job as an electrician in the Detroit area.

George listened and then said, "There is no way to bring charges against this Detroit man. It would be practically impossible, because there was no jurisdiction covering domestic cases between the United States and Canada."

After we left the hospital, George and I were sitting down having a beer in the Norton Palmer Hotel.

George said, "What do you think of all that, Pete?"

"Well," I said, "what goes around comes around."

We both agreed that after what she did to Norm, it was too bad she didn't die herself from the overdose.

George said, "She's a no good son of a bitch, and I'm going to stick with what I told her—there's nothing that can be done to collect any of the monies that she had loaned her boyfriend, and she is completely out of luck."

It turned out that Diane's mother lived on the outskirts of Windsor, and Diane was able to move back home and live with her. I never again saw or heard anything from or about her.

Dick Hymen

A remittance man was someone who came from a well-to-do family in England who was sent to Canada because they were difficult or the black sheep of a family. They would be sent a monthly remittance or allowance to keep them here. The families didn't want to have to deal with these particular family members on a daily basis.

I always had the impression that Dick Hymen was one of these remittance men. He was also one of the worst cowards I have ever known.

Dick was a writer for the *Hamilton Spectator* newspaper before he came to Windsor. After moving here, he became one of the offbeat members of our club at Ma Ellis' on Goyeau Street. To supplement his monthly remittance, Dick wrote booklets he called "The Voice of Courage." He then sold them to funeral homes to give to the people in mourning to help them overcome their grief and sorrow. The booklets were well received.

Dick did research at the library to help him compose the booklets. He was very successful and also did public speaking for different organizations like the Kiwanis Club that would engage him to give speeches about philosophy on certain occasions. He was quite popular.

However, when I got to know him better, I discovered that he had less courage than anyone I knew or had ever met. If Dick owed any bar or Ma Ellis money, he would go into hiding until he came up with the money to pay them.

One day Dick was with me when I had go to Detroit to pick up some business papers and do a couple of errands while I was there. When we were about half way through the Detroit Tunnel, he suddenly got quite agitated. He said, "I can't go into the US because I'm wanted for treason."

"What in the world are you talking about?"

"Well, it has to do with a telegram I sent to the president of the United States."

That was all he ever told me about the telegram. When we got to the US customs, I told the officer that I had forgotten some important papers and that I needed to go back to Windsor to pick them up. He let us turn around with no questions asked, and back we went.

Back in Windsor, we were going to the British American Hotel for lunch and a beer, but Dick said he couldn't go in there because he owned some money to the bartender. As we were going to another location, Dick again said he couldn't go there either because he owned them money as well. I can't remember if we ever found a place to go for lunch and a beer that day.

A few weeks later, Dick came into the club looking for George Yates, my lawyer. He said, "I need to see him and need him right away on a serious matter. I'm in a whole lot of trouble."

I said, "George is here, but he's in the bathroom, so you might as well sit down and wait for him."

When George came out, Dick said to him, "I don't know what I'm going to do, because I'm about to be charged with murder."

"What are you talking about and who are you supposed to have killed?"

"I just heard on the radio that Norm Riordon shot and killed himself in his apartment on Pitt Street."

"Dick, did you shoot the poor bugger?"

"No, but I've got all his clothes on."

"Well, did you steal them?"

"No."

"Okay, then, why do you think you are in trouble?"

"Norm and I were out drinking last night, and he asked me to come up to his apartment with him. When we got there, Norm said he had clothes that he wouldn't be using, and he wanted me to have whatever I could use that fit me."

He went on, "I tried on one of Norm's suits, shirts, and a top coat. They fit me perfectly, and Norm told me to keep them. I thanked him and told him how much I appreciated him giving them to me. So I took the clothes and left Norm's apartment about midnight and went home. When I got up this morning, I put on some of Norm's clothes, and when I got into my car, it came over the radio that Norm had shot himself."

Dick yelled at George, "Look at me. Norm Riordon is dead, and here I am with his clothes on. Everything I'm wearing today belonged to Norm. I expect to be charged with murder at any time."

"Well, if you didn't do it—and I don't think you'd have the guts to do it—there should be no problem."

"I'm still a physical wreck and I'm afraid I'll be arrested and I just don't know what to do."

George was losing patience with Dick and finally told him, "Just go home. Go back to bed and stay there for three days. If you hear anything, contact me, but if you don't hear anything after three days, just go about your normal business."

It appeared that Dick followed George's advice, because no one saw hide nor hair of him for about a month.

Peter James McLean

Lawyer/Suicide

A group of us used to meet for coffee in the morning at the Honey Dew Restaurant on Ouellette Avenue in downtown Windsor. We would just sit and discuss the news or whatever topic someone came up with. The group was made up of lawyers from the Canada Building.

There was an advertising man named Bob Dearth in Detroit who was interested in my mining properties, and one of the lawyers in our group said he was a friend of Bob's. He began asking me different things about Bancroft, on behalf of Bob Dearth, to check me out and find out how everything was going. We had a nice discussion for about ten or fifteen minutes before he said he had to leave and go up to his office.

Within an hour, we heard that this lawyer attempted to commit suicide. He had shot himself with a WWII .45-caliber revolver that contained its original bullets. Because of the age of the gun and bullets, it had lost a lot of power, and the bullet didn't penetrate his skull. He was injured, though, and it knocked him out cold.

It was a terrible thing to have happen. I couldn't understand how such a nice, cool, and collected person could have so many problems on his mind that were worth committing suicide over. It turned out he had been doing quite a lot of gambling in Detroit and had used up all of his client's trust money.

Magistrate McDonald of Windsor wanted to make a name for himself and charged the lawyer with attempted suicide. He did it for personal reasons that had nothing to do with the suicide attempt. The results were that the lawyer lost his license and his large law practice.

The last I heard of him, he was working on a farm near Tilbury, Ontario. George Yates had to take over his practice and straighten it out.

Johnny Jones

I'm adding this story here because had he lived in Windsor, this friend would have definitely been a member of our club.

I had a friend in Cobalt, Ontario, by the name of Johnny Jones. He was a prospector and was very wealthy. He had found a large silver deposit in Kidd Township just outside of Timmons, Ontario, during the mid-1960s. He had a hard time getting a mining company interested, because the silver was located on open pastureland and the area was very accessible to anyone.

One day at home in Windsor, I got a phone call early in the morning, and this voice said, "Guess who it is. It's me, Johnny Jones, Peter, and I'm in the Elmwood Hotel, and I've got my new wife with me. I want you to come out and meet her. Come by this morning if you can make it."

I said, "Sure, I'll be there within the hour."

So I went to the Elmwood and met his new wife. She was a dyed blonde in her early thirties, while Johnny was a fellow in his middle seventies who looked every bit his age. He was stooped and looked like an old-time prospector. At first, she gave me the impression that she was a gold digger. However, when I got to know the lady, I realized she was a very respectable woman and that Johnny had found himself a very good wife.

After the pleasantries, Johnny said, "Pete, what I'd like you to do since we're on our honeymoon is drive us to Detroit and get us checked into one of the best hotels over there. I want to take my wife shopping and sightseeing. I want to be downtown in order to do some business that I hope will work out with some people from the Grosse Point area."

I just said, "Okay, Johnny, let's go."

Johnny had never been to Detroit, and he was combining his honeymoon with business.

Johnny had a brand-new Buick Roadmaster station wagon that was sitting very low to the ground, and I asked, "What in the world do you have in this car?"

He opened the trunk and said, "I've got it loaded with copper ore that I need to show some people."

It looked like there was close to a ton of ore in the back of his brand new car.

We left the Elmwood Hotel and went through the tunnel to Detroit. The customs agent asked me who owned the car, and I explained to him that Johnny owned the car and that I was just driving him and his new wife to the Sheraton-Cadillac Hotel. The agent asked Johnny for the ownership papers.

Johnny didn't have those or any other papers for the car, which he had just bought and paid cash for from a dealership in Cobalt, Ontario.

The agent then asked Johnny if he could see his driver's License or anything else he had for identification, but again Johnny didn't have anything on him. The agent kept smiling and was trying to figure out what was going on, because Johnny is a short, fat, real rough-looking character.

I told the agent, "The people up north do everything by word of mouth and a handshake. That's the way business is done up there, because some people can't read or write. The handshake was law."

The agent said, "I want to look in the back of the car."

When it was opened, he couldn't believe what he was seeing. Johnny told him, "It's filled with copper ore from a rich claim I found, and I'm taking it to show some people in the Grosse Pointe area."

The agent didn't know what to do with us. Johnny was surprised the agent didn't know who he was, and he said to the agent, "I'm the Cobalt Silver King."

The Agent didn't know what that was supposed to mean, and Johnny told him, "Everybody knows who I am, and I'm a good friend of Sir Harry Oakes." He figured the agent would know who he was.

Then Johnny thought of something. He rummaged through the car until he found what he was looking for. It was a copy of the Cobalt newspaper, and there on the front page was a picture showing Johnny Jones, Silver King of Cobalt. He had been dressed up and was at the head of a parade.

As a last resort, the agent asked, "Do you have a credit card?"

Johnny replied, "No, I don't use one."

He then pulled out a roll of bills as big as your fist.

The Agent finally said, "Okay, you can go. Have a good time and enjoy yourself. But be very careful showing your money around strangers in the Detroit area. I hope your honeymoon and marriage work out for you."

I'm sure it was one of the strangest situations the agent had ever come across. I cannot begin to imagine how this would have worked in today's Homeland Security era.

I drove Johnny and his wife to the Sheridan-Cadillac Hotel and got them checked into the bridal suite. I told the hotel manager that Mr. Jones and his wife were from Cobalt in Northern Ontario and asked if he would look after them. Also, I asked him to make sure they had a chauffeured limousine so that they were protected when sightseeing.

I came back home to Windsor, and about a week later, I got a call from Johnny telling me he had just flown into Toronto from Detroit. He asked if I would pick up his car at the Hotel in Detroit and bring it to his new home on Avenue Road in Toronto.

I picked up the car, and within a couple of days, I had the car in Toronto.

When I asked him how he made out with his copper claims and everything, he said, "Terrific, terrific, everything is going to work out. Texas Gulf is doing aerial geophysical electromagnetic surveys over the property now, and if the results turn out good, I'm all set again."

The results of the Texas Gulf survey were very positive, and the property turned out to be a very large producing mine. It started a big rush in Timmons in 1964, the largest since the uranium boom of the 1950s.

At one time, Johnny had worked with Sir Harry Oakes, who made his millions from his gold discoveries when he was a prospector in the Porcupine area of Northern Ontario. He owned the second largest gold mine in North America around 1920. It was said that he made his initial gold strike when the Chinese man who owned a hotel and restaurant tried to collect money he was due. Sir Harry threw an axe at him, and it hit an outcropping of rock and broke off a piece, uncovering a rich gold vein. This was the beginning of the big gold boom in Northern Ontario in the Porcupine, Kirkland Lake, and Timmons areas. Johnny was in the vicinity at the time and said the story was true and that that was exactly how it happened.

Sir Harry Oakes was mysteriously murdered in the Bahamas in the 1940s, and the case was never solved.

Although the stories of finding gold by throwing an axe or hammer or slipping on something sound far-fetched, I tend to believe they are true.

TREASURE HUNTING

In the early part of the 1950s, during my research, I came across an article in the library about all the numerous ships that had been sunk in the Great Lakes. From what I understood, there were no treasure hunters at that time looking for these ships to recover the cargoes.

I studied everything I could find on sunken ships in the Great Lakes, and after a great deal of time to research, I put together a treasure map and folder. When it was finished, I felt it would be an excellent mail order sales item, and I named the project Guy Underwater Exploration, Ltd.

Through classified ads I ran in the *Mechanix Illustrated*, *Popular Mechanics*, *Popular Science*, and *Sports Illustrated* magazines, I sold the maps and folders for two dollars each.

I rented a mailbox at the Sandwich Post Office in Windsor. The box number was 7174, and I kept and used it for years and years.

As I had expected, the treasure folder was a very popular item, and I sold them through the mail by the thousands. They made me realize that treasure hunting, especially in the Great Lakes, would be a very popular project. There was more interest in treasure hunting than in any other subject I had ever worked on.

I found most people never realized the number of shipwrecks there were in the Great Lakes or the number of lives lost in those wrecks. The Great Lakes border eight states—Illinois, Indiana, Michigan, Minnesota, New York, Ohio, Pennsylvania, Wisconsin—and the Canadian province of Ontario. The Great Lakes treaties between Canada and the United States also include the province of Quebec.

The Great Lakes are busy commercial waterways that can be churned up into deadly storms. It can be very treacherous, and thousands of ships, crews, and passengers have been lost. Today, most people only relate shipwrecks to the fate of the *Edmund Fitzgerald*, which went down in Lake Superior on November 10, 1975, with twenty-nine men on board, because of the song Gordon Lightfoot wrote and recorded about it.

Inventing the Waterscope

I had been selling the treasure folders for well over a year, and one day, Ted Boomer and I were discussing it during one of our beer drinking episodes in our favorite tavern. I said, "Wouldn't it be great if I built an inverted periscope with lights and magnifying glass at the end of the periscope that we could use in order to search the bottom of Lake Erie and Lake Huron? Lake Erie seems to have the majority of sunken ships in it, and it is the shallowest of the Great Lakes."

Ted agreed with me, and I was sure I could build such an apparatus. After I got started and built a complete prototype with the light, tubes, and glass, I called it a Waterscope and had it patented. We were testing it in the Detroit River, and it worked great.

We found an old sunken car that was loaded with a rumrunner's liquor bottles, which we salvaged for ourselves. We felt this experiment was highly successful. The Waterscope was going to work, and now we needed a ship to mount it onto so we could start our treasure hunting in earnest.

The Waterscope was such a new and novel idea that we received a lot of publicity from it. There was also a picture and a write-up of it in *Mechanix Illustrated* magazine.

We now had a Waterscope that worked just as we had envisioned it would, but we still had no ship to mount it on. A well-known ex-Great Lakes captain by the name of Spears was very enthusiastic about the invention and our ideas for using it.

I have to note here that there was one thing that was outlandish about Captain Spears, and that was that he lived in the hull of an ex-Great Lake's freighter in the city dump here in Windsor. Ted and I had

Peter James McLean

some great times visiting and discussing plans with him in his unusual abode.

When taking a photograph of the Waterscope with Ted and me along with Captain Spears, we had to turn the captain a certain way for the camera because of the swelling of his testicles. We never learned what caused it, except that he told us he had a medical problem.

Captain Spears was so excited about the Waterscope that there was a picture and accompanying article in the *Windsor Star* newspaper at the time. That was all well and good, but we still needed a ship in order to start our treasure hunting in the Great Lakes.

Newspaper article on my Waterscope Invention
Printed with permission from The Windsor Star which retains copyright

A Ship Is Found

The US *Chaparral*, an ex-US icebreaker, was going to be refitted and used as a Pelee Island Ferry. The ship was bought in 1946 by Vince Barrie, whom I personally knew. Vince was the mayor of St. Thomas from 1959 to 1964 and again in 1967–1968. After some negotiations, Vince agreed to sell me the ship.

The ship had a very interesting history. It started out with the Walkerville Ferry Company and ran between Walkerville, Ontario, and Detroit when it was named the *Halcyon*. Walkerville later became part of Windsor. During WWII, the US Coast Guard made her into an icebreaker and buoy tender and changed her name to the US *Chaparral*. When I bought her, I renamed her the *Jolly Roger*.

The icebreaker was a large ship approximately four hundred feet long with a wide beam. It was made of steel and metal from top to bottom with no wood used anywhere on the ship. It was docked in Port Stanley, Ontario, and I had to have it towed to Windsor by Captain McQueen of McQueen Ltd. in Amherstburg.

It was quite an adventure towing the boat to Windsor.

There were three people with me when I drove my car to Port Stanley to pick up the ship. Two were workmen I had hired, and the other person was Ralph Smith, a wealthy farmer from Essex, Ontario. Now that my car was in Port Stanley, we had to get it onboard the ship for the return trip. I don't remember exactly how we did it, but I think we used planks to make a ramp, and the Cadillac was driven at a high rate of speed up the ramp and jumped onboard. It had to have been done that way, because there was no tow truck or other lifting equipment near us.

Captain McQueen and his tug the *Atomic* arrived in Port Stanley to start the trip at around nine in the morning. I sent the two boys to get two loaves of bread, a pound of baloney, ketchup, a jar of peanut butter, and between four and six cases of beer. We discovered early that we had lots of beer to drink, but very little food.

One of my workers apparently had too much to drink and decided to dive over the side of the ship and go for a swim. The problem, however, was that the tug was pulling the ship much faster than he could swim. We ended up having to shut down the tug and go back and rescue him. Fortunately, everything turned out just fine. But there was so little food on board for all of us that we did more drinking than eating. We were all looking pretty rough when we finally got to Windsor.

Towing the ship and docking it created so much excitement that the *Windsor Star* took a picture of Ralph Smith, Captain McQueen, and me on the ship and wrote a story on it. People were curious to know where the ship had come from and what it was going to be used for.

I had to hire a tow truck to lift my car off the ship. The tow truck driver couldn't understand how we managed to get the car on the ship in the first place. He had to lift the car up in the air and then set it back down on the dock.

Before bringing the ship to Windsor, I had rented the ferry docks and ferry building at the foot of Ouellette Avenue in Windsor. It was an ideal location because it was in plain view of downtown Detroit. An old receipt I have says I paid the city of Windsor thirty-five dollars per month for the rental of this great location.

Treasure Unlimited

After getting the ship docked in Windsor, one of the first things I did was to have it repainted its original black color with a white superstructure. This was needed to cover up all the rust spots and to make the ship—from the outside, at least—look brand-new. On the Detroit side of the ship I had the name "Treasure Unlimited" printed in big letters that ran the length of the ship. This was done for advertising purposes so that the Americans in downtown Detroit could easily see and read it.

I had the captain's cabin and the upper deck portion of the ship nicely furnished. The meetings and discussions I had with investors were all conducted in this area, and it gave them a positive read on the project.

The "Treasure Unlimited" name created so much curiosity and interest that Americans came in droves to tour the ship. Not only were people interested in the ship, but there were a lot who were interested in investing in the project. This made it unnecessary to do any other advertising. We had enough investors and lots of money in the treasury to keep the project going. We got more publicity from the July 1953 edition of *Inside Michigan*, which ran an article mentioning that I had done considerable research on this project. You couldn't buy the type of publicity we received.

With everything on the ship, investors could come aboard for a tour and examine the Waterscope, which I had mounted on the stern.

People could look through it and see the bottom of the Detroit River. The investors realized this was going to be a successful operation.

Treasure Ship at Ferry Dock in Downtown Windsor
"Treasure" decal positioned towards Detroit, Michigan

R. M. (Dick) Harrison

Dick had been a reporter for a newspaper in Winnipeg, Manitoba. According to him, he had known Mrs. Adie Knox Herman, who owned the *Windsor Star* newspaper. Mrs. Herman offered Dick a job for life if he would accept a position with the *Windsor Star*. Dick readily agreed and made the move.

Dick wrote a newspaper column called *Now* and became well known and respected. His column became a must-read to know what was happening in and around our town. He soon became a patron of our club at Ma Ellis'.

Dick was very interested in the treasure ship and the programs and activities I was involved in. Dick had a prominent name in Windsor, and because of this and his interest in the Treasure Ship, I asked him to be on the board of directors of my Guy Underwater Exploration Company, which included the treasure ship and all its activities. Dick became the company's vice president.

Mrs. Herman allowed Dick to be very independent, and he could do the writing of his column and all the other things he did whichever way he wanted to. He had quite a knack for finding out information before it was publicized, and he would include this information in his column. Dick said a lot of the information for his column in the paper came in as tips over the telephone.

Dick was a very eccentric individual, and claimed that every night after midnight, he would write his column. He said he never did any writing before that time. He also claimed that his office at the newspaper had a padlock on the door, and he would go in and out of his office through a window. Everyone wondered how he could get away with such outrageous behavior, and no one knew the reasoning behind his mysterious activities. But apparently his idiosyncrasies were accepted, because Mrs. Herman never questioned them.

Dick was also a heavy drinker. He used to condemn drunk drivers in his column, and I often said to him, "Why are you so hard on drunk drivers?"

"It's just something that's very appealing to the public; they like what I write in my columns emphasizing that drinking and driving don't mix."

Dick didn't drive a car. I never asked him whether he lost his license through drinking and driving, so I wasn't sure of the real reason he didn't drive or own a car.

While Dick was with me on the treasure ship adventure, with the different things he needed to do, I became his willing driver.

I had hired a man named Sam Baker to be the guard on the treasure ship, mostly for looks. We had Sam wear a uniform, and part of the uniform was a .45 revolver that was strapped to his hip. It made him look the part of being a well-armed guard. However, Sam was out of service on the railroad because he had Saint Vitus' dance (a name we used at that time for anyone with Parkinson's). There were times he was shaking so badly I would have to roll his cigarettes for him.

I had written to the Windsor chief of police asking for a permit for Sam to carry the gun, and the chief sent an officer down to interview

Confessions of an Eccentric Dreamer

Sam. He reported back to the chief that there was no way they would give him permission to carry a gun.

One day in the office, when Dick, our secretary, Maria, and I were all present, I told Dick we had to get a gun permit for Sam, but the chief of police had turned us down. Dick said that would be easy to do and that he would take care of it. He said he would write a personal letter to the chief.

Dick started dictating the letter to the secretary, but he was so intoxicated at the time that he was staggering all over the office, and Maria couldn't keep up with him or understand what he was saying. He finally told her to get out of his way, and he sat down and typed the letter himself. The typewriter just sang as he was typing so fast. He pulled the sheet of paper out and said it would work. The letter was perfect. It was one of the nicest letters I have ever read, and I told him so. I took his letter and immediately mailed it, and within days we had a permit for Sam to carry a gun.

The reason I mention this is because with Dick's heavy drinking, it illustrated how he was able to write such a good newspaper column. Whether he was intoxicated or suffering through a bad hangover, nothing affected the caliber of his writing.

One morning, Dick and I were together at our favorite club at Ma Ellis' when the police came in and arrested Ma for running a blind pig. Dick and I were charged with "found in." We left and went to another blind pig we knew on Windsor Avenue. We were only there about an hour when the police raided this place and again charged Dick and me with being "found in" and charged the owners for running a blind pig.

It was about noon, and the hotels were now open, so Dick and I went to the Fisherman's Cove in the Prince Edward Hotel to have a few drinks in peace.

I commented, "Our names are probably going to appear in tomorrow's newspaper."

Dick said, "Don't worry about it. No one is going to write an article with our names in it."

Dick was correct in that reporters were so clannish that they would not write anything detrimental concerning another reporter.

These are just some of the highlights of my association with Dick Harrison. A couple of years later, Dick was flying back from Toronto and passed out on the airplane. They had to divert the flight and had an ambulance waiting to take him to the hospital. The doctors examined him and told him that if he didn't stop drinking completely, his days were numbered. They scared Dick so bad that he never touched another drink.

I was always of the opinion that quitting drinking so abruptly was a severe shock to his system. It was so severe, in fact, that Dick passed away approximately six months later.

During the time I associated with Dick Harrison, we forged a good friendship, and I missed him after his death.

R. M. (Dick) Harrison
Printed with permission from The Windsor Star which retains copyright

Vince Meli

One of the potential investors that I recall was a very nice, college-educated young man from Detroit by the name of Vince Meli. He and I got to be quite friendly, and one day I invited him to our apartment to have lunch with Marion and me. Vince was very happy to accept my invitation, and the day that he came, he had a large gentleman with him.

I asked Vince, "Is he a friend of yours?"

"No, he's kind of my bodyguard, but he won't interfere with our visit or meeting."

During lunch, the "bodyguard" was very polite and quiet to the extent that he was not involved in any of our conversations. After lunch, I had Marion take a picture of Vince, the "bodyguard," and me on the balcony of our apartment. I then took a picture of Marion with Vince and his "bodyguard."

Our get-together was very pleasant, and we all enjoyed it. Vince asked me if I would come to Detroit and meet his father. He said he was very interested and that his dad would like to invest a large amount of money in the treasure ship.

I told Vince I would be glad to meet his father and present all the plans that I had for the treasure ship and the underwater explorations that I was going to be doing. Vince gave me an address of 4000 and something Gratiot Avenue, but he didn't say whether it was a house, building, or whatever. He only gave me the address.

About two days later, we set a time and date for me to come to Detroit to meet Vince's father. When I got to the address, it turned out to be an empty store with curtains on the windows that you couldn't see through. Putting that together with the "bodyguard," I had the impression that I was walking into a mafia meeting room. I knocked on the door, and when a man answered, I told him who I was. "Come right in," he said. "Mr. Meli is expecting you."

I explained to Mr. Meli that the ship wasn't ready and that I had to spend quite a sum of money to repair the boilers and would rather that he waited a few months before investing any money in the project. I soft-pedaled my sales pitch with Mr. Meli, because having watched so many gangster movies, I felt I was dealing with a mobster. I did not want any money from Vince or his father. I thought I would be involved with the mafia, and there was a good chance of being killed if they thought I had crossed them.

Peter James McLean

After the visit, Vince called on the telephone, and I told him that things were not progressing the way I would like them to and that I would contact him later. That was my last contact with Vince.

I found out later that Angelo Meli played a leadership role in the Detroit mafia until his death in 1969 at the age of 67. He had a nephew by the name of Vincent Meli (Little Vince) who had graduated from Notre Dame in 1942. He became the first college graduate of the second generation of Detroit underworld figures.

Is the Vince Meli I met part of the Detroit mafia, or was I reading too much into it?

The "Bodyguard", Vince Meli and Marion

Ulysses Lauzon

Our unofficial headquarters for the treasure hunting was at the British American Hotel, which was practically right next to the ship at the corner of Riverside Drive and Ouellette Avenue. We even had a telephone line between the ship and the hotel to keep us closely connected.

Bars are a font of interesting information and this one was no exception

One day talking with an off duty detective sitting at the bar, we got into a conversation about my previous projects. I related how I had worked with Peter Hedgewick and Roger Lauzon at International Tool and Die here in Windsor to make the plastic injection molds for the toy guns I had invented earlier. International was a successful company that had been the first to invent safety caps for medicine bottles.

The detective then brought up the subject of prison breaks and a bank robber name Ulysses Lauzon who hailed from Windsor.

In August, 1947, Ulysses along with two other inmates, Nicholas Minnelli and Mickey MacDonald escaped from Kingston Penitentiary. At that time, this was the second break out at this maximum security prison. Minnelli had been captured fairly quickly. However, Ulysses and Mickey were linked to a bank robbery here in Windsor where the robbers got away with about $40,000. These two were described as being vicious and ruthless criminals. Through Lauzon they were thought to have connections here. Lauzon also had an ex-wife living here in town and it was felt they could find a safe place to hide out. A manhunt ensued but they were not found.

In July, 1948 Lauzon's body was found badly beaten and shot execution style in a swamp in Mississippi. There were rumors that MacDonald was killed at the same time as Lauzon but his body was never found and his whereabouts remain unknown.

Since I didn't get transferred to Windsor until 1948, I was not really familiar with this story and its connection to Windsor.

The detective then proceeded to tell me that Roger Lauzon at International Tool was the brother of Ulysses Lauzon, the bank robber. He said that over the years some people speculated that Roger got money from his brother's bank robberies to help start International Tool and Die.

I could not believe what I was hearing. I had found Roger to be an outstanding individual to deal with. Anyone who knew him would not believe any of the rumors floating around at that time. Roger was an upstanding, honest person who was well-liked.

Underwater Salvage & Diving Inc.

At this particular point in time, there were no scuba diving schools because the necessary equipment was not yet available. Diving had to be done by deep-sea divers with their helmets and heavy equipment.

We needed an experienced deep-sea diver available for our underwater salvage operations. Because of this need, I started a diving school and hired a man named Hub Moore as the instructor. Hub had twenty-plus years of experience. The diving school would operate on our treasure ship.

I had known Hub as a neighbor of a Mrs. Gould, who ran a blind pig in the Remington Park area of Windsor. I found out he had extensive diving experience and had done a lot of diving on the Great Lakes for the Royal Canadian Mounted Police. He was well liked and respected for his work and experience. I felt fortunate to have him on the payroll.

It was important that Hub accompany me whenever I went looking for the necessary diving equipment we needed.

One day I learned that a Mrs. Dodge in Detroit had some diving equipment for sale. I called, and she asked me to come on over and check it out. It turned out that Mrs. Dodge lived in a mansion on Jefferson Boulevard in Grosse Pointe, Michigan, and was part of the automotive family.

She was very friendly and obliging about showing us the equipment that had belonged to her late husband. With Hub's approval, I struck a deal with Mrs. Dodge and purchased all the equipment she had.

When we finished, I told Hub we would stop at a bar and celebrate. Hub said he couldn't stay long because he had to get home to take his "old lady" dancing. After a couple of drinks, we came back to Windsor, and I drove him home. He asked me to come in and have one more drink with him and meet his "old lady." Curious about his wife, I agreed and went with him. To my complete surprise, his "old lady" turned out to be his mother, and he was actually going to take her to a dance!

What puzzled me was how a forty-year-old masculine man could be such a mama's boy. I never did figure him out or come to understand his relationship with his mother.

Another diver we had was Avery Hampton, who was from the state of Georgia. One day we were driving in Avery's car on Gratiot Avenue in Detroit when at a stoplight there was a black man driving the car beside us in the curb lane. One look at him and Avery went crazy.

He said, "Look at that son of a bitch. If I was home in Georgia, I'd just shoot the bastard. I'm going to fix him."

He then proceeded to crowd the man's car until he forced it right into the side of a parked car. Avery then told me, "Blacks are hated where I come from."

Avery himself was a dyed blond who had been in some movies and thought himself better than most people. Thinking about it, I realized the whole incident with the black driver was because he was driving a brand-new car while Avery was driving a much older one. That was the real reason he hated this man so much.

One time he invited Marion and me to his home. It turned out that practically everything in the house from carpeting to furniture was covered in plastic sheets. It was extremely cold and uninviting.

Both Hub and Avery were experienced divers and worked for me during the time I had the treasure ship.

Each class we ran had a maximum of twenty students. Classroom work included the use of special underwater welding and cutting equipment, because these skills are essential in actual ship salvaging.

While operating the deep-sea diving school, I had an office in the First National Bank Building in downtown Detroit. I was using it to enroll students for the classes.

One day I received a letter from an ex-student who had gone to Saudi Arabia to work in the oil fields. It was the most threatening, abusive, and slanderous letter I had ever received. I thought I had better take it and show it to my lawyer, Nobel Lawson. I wondered if it was against the law to send such a missile through the mail.

On my way over to Nobel's office in the Buhl Building, I had a remote sense that I was being followed. But I could think of no reason why someone would be following me.

When I arrived at Nobel's office and spoke to the secretary, she said, "Peter, you'll have to wait a few minutes because Mr. Lawson is busy on the telephone. Just have a seat, and he won't be long."

I had been sitting there for about two minutes when a man came in the office and spoke to the secretary. He said, "I'm from the FBI," and showed his credentials.

"What I would like to do is to talk to Peter McLean and his attorney if I could."

I went into Nobel's office first and said to him, "There's an FBI agent outside wanting to talk to you in regard to me."

Nobel called his secretary and said, "Send the FBI man right in."

The FBI man came in and sat down. Before he could say anything, I told Nobel, "I want you to first see the letter that I just received from an ex-student."

The FBI agent started to laugh and said, "That's the reason I'm here. Your ex-student has also written the same type of letter to the FBI."

I explained during our meeting that I knew the reason behind the ex-student's violent letters. The student had joined the diving school and attended two instruction classes. He then came to me and said, "I've got a new job, and I want to take it, and I want my money back from the school."

All the students paid for the class in full in advance and were told there would be no refunds and the contract was binding. I'd told him, "If you want to quit the school, that's fine, but I'm not going to return the money that you prepaid. You were told in the beginning that there would be no refunds."

"He got into quite a huff, but there wasn't too much of an argument." I continued, "Before he left, though, I again emphasized that the classes were paid in advance and that there would be no refunds because of the cost and expense of operating the classes. He then left, and I never gave him another thought until I received this letter."

I said to Nobel, "If anything happens to me, you'll know who to look for."

He and the FBI agent agreed, and Nobel added, "After receiving a letter like this, you should keep in mind that there are people out there who can quickly turn against you."

The FBI man finally said, "Mr. Lawson, I now understand the reason for the letter and the circumstances surrounding it. The FBI is perfectly satisfied that there was no wrongdoing on Mr. McLean's part."

He also told me, "I've been following you for several days, but it was standard procedure when we're trying to assess this type of situation. Thank you, gentlemen."

Then he left.

When the FBI agent was gone, Nobel said, "Just forget about the letter. There's no money in suing or taking any action against this ex-student."

That was the end of that little episode. I told Nobel to keep the letter in his files for future reference. I should have asked for a copy of the letter at the time, but I didn't.

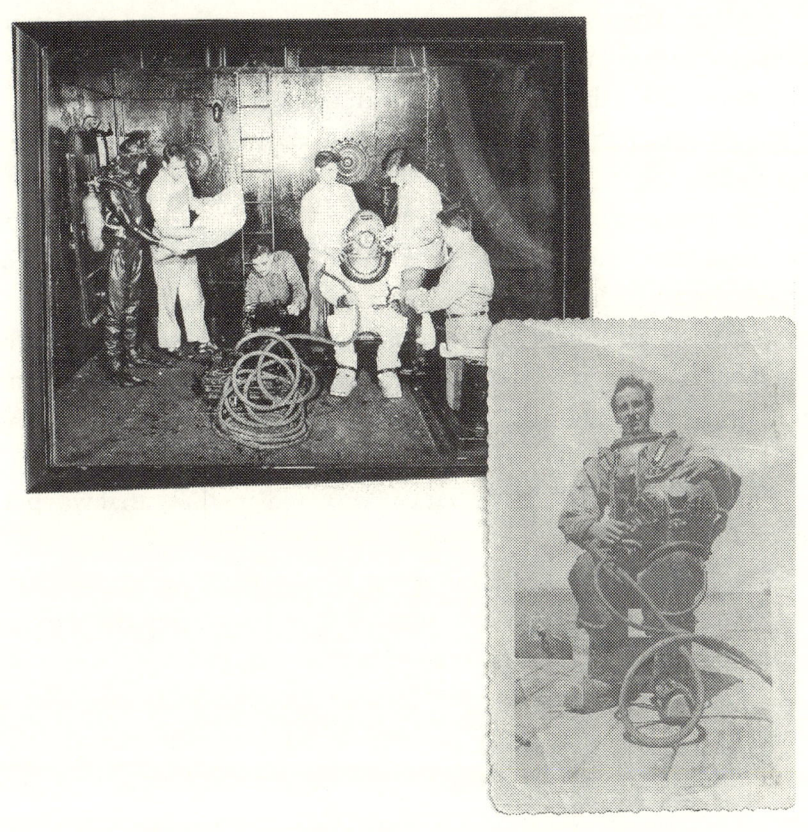

Underwater Salvage & Diving, Inc. Diving Class
Avery Hampton

The Real Treasure

During this time I had lots of visitors and interest from people in the Detroit area who wanted to invest in the Great Lakes underwater treasure expeditions.

Everything was going well financially, and I began to make plans to get the ship into operation. The first thing I found out was that I had to have a licensed captain and engineer onboard the ship at all times in order for me to get it operational. The ship was steam powered with big boilers in the engine room. It was also going to be necessary for me to have the boilers rebuilt.

All this had to be done in accordance with the rules and regulations of the federal government. After four years of doing all the preliminaries, I began to realize that the expense of all this was becoming prohibitive when I factored in the bills for dockage fees, the hydro, and maintenance of the ship.

The board of directors consisted of Ralph Smith as president; Dick Harrison, my newspaper writer friend, as vice president; and me as secretary. This meant that through the board of directors, I had complete authority and was able to proceed with whatever projects I felt would help the operation. I knew that for us to succeed with our treasure hunting and exploration activities, we would need a smaller ship—a ship that could be operated with a minimal crew.

I looked for a smaller ship but couldn't find anything that was both affordable and workable for our needs.

I told Captain McQueen that our bills were piling up and that I wanted to know the best way to proceed in order to scrap the ship and go on from there. The captain suggested we strip the ship down to the hull. Then that portion of it could be used as an excellent barge because of the heavy-duty strength of the ship. I had the captain tow the ship to his shipyard in Amherstburg to start the dismantling process. The loss of our treasure ship was a real heartbreaker for me. But I also knew it had to be done.

McQueen went to work on the ship, and later I got a call from him. He wanted me to come down to the shipyard because he had a surprise for me. When I got there, I found out that the ship from the engine room floor to the bottom of the hull was filled with solid lead. At that time, lead was worth approximately twenty-five cents per pound, and in the hull of the ship we had tons of lead for salvage.

It worked out that our "treasure" was in the ship itself and that we had been walking around on it all this time. When they built icebreakers at that time, the ships were built with heavy steel hulls. They were not made to break through the ice with the bow, but to run up onto the ice, and when coming down, the weight of the ship would break solid ice, which was four or five feet thick. The hull was built, and then lead was poured into it from stem to stern.

At that point I put an end to the project. The money made from scrapping the ship and selling the hull with all the lead in it was enough to pay off all debts and monies owed. There were no repercussions from anyone, anywhere.

There was enough money left over to start a new adventure.

NOT MY FINEST MOMENTS

Early Impaired Driving Conviction

Late one summer night, I was returning home from the Fisherman's Cove Tavern and was driving along Victoria Street behind a squad car. All of a sudden, the squad car slammed on its brakes, and I smashed into the rear of the police car. It certainly caused a lot of excitement for the officers.

At that time they didn't have Breathalyzer tests, but they judged your condition by the smell of your breath and your walking and talking ability. They said I was drunk, and they arrested me and charged me with drunk driving.

The reason the squad car had stopped so suddenly was that someone had broken the front door of the newspaper building just as the squad car was passing by. That sudden stop was bad luck on my part.

Just before I left Fisherman's Cove, the conversation got around to drunk driving and getting arrested for it. I sat there and bragged about never having been arrested and not expecting to be in the future. I discovered quickly that when you brag about these things, it will come back and bite you in the ass.

At that time, Magistrate Hanrahan had vowed that all drunk drivers were going to jail, and he had been practicing this for about six months. I was caught in his web. During the trial with the evidence that was given I was convicted of drunk driving and sentenced to seven days in jail.

The time for me to go to jail wasn't specified, but when I talked it over with Dick Harrison, my writer friend, he said things were fixed and arranged so that if you report to jail on a Monday before seven in the evening, your time will be served early the following Saturday. I realized this was an easy way in and out, because you would only be there from Monday evening to early Saturday morning.

That Monday, I decided I might as well serve the jail term, so I went and parked my car at a friend's house and told Marion I was going up north and would be gone a full week. I felt I had everything arranged so that while I was in jail, everything would be taken care of, and nobody would know where I'd been for the week.

After taking care of the car, I went down to our club on Goyeau Street and started to drink with the boys. I had it in the back of my mind that I had all day, as long as I hit the jailhouse by seven o'clock.

At six thirty I called a taxi and told the driver to take me to the county jail in Windsor. The driver was very leery, what with my being drunk, and he didn't believe what I was saying. I was very emphatic, paid him twenty dollars, and insisted that he take me right to the front door of the jailhouse.

When we arrived at the jail, the taxi driver rang the bell and told the guard there was a drunk in his cab who wanted to come into the jail. The guard said, "Like hell, we've got enough drunks in here now, and the drunk in your cab must be thinking of a free night's lodging."

I ended up going to the door myself and told the guard what I was doing there and that I had to serve the seven-day sentence that Magistrate Hanrahan had given me. They didn't believe a word I said, and they told me to wait outside while they went to check their admittance papers at the jail. They checked but they couldn't find any record of me.

Again they came back and told me that they had no record of me, but I kept insisting that they check again. They finally said they would have to phone down to city hall and see if there was a record of me down there. All this time, they kept me outside sitting in the taxicab. They were bound and bent not to let me in the jail.

After nearly an hour, the guard came back and said, "Okay, McLean, you can come on in, because we now have confirmation of your sentence."

It was getting close to eight before I finally got booked into the jail. The guards were talking, and they said that they had never known of a drunk trying to get into the jail to serve his sentence. I finally got admitted, and they put me in the hospital portion of the jail to sleep it off.

Being an hour late getting into the jail worried me, because I wondered if I would be credited for the time on Monday. They told me later in the week that they would allow me time served for that Monday.

During the week in jail, I spent the time delivering meals to the other prisoners, and it seemed like a good job. One prisoner in there at the time was a man who called himself *Yorky*. He boasted that he had been arrested and jailed over four hundred times because of his drinking. Yorky was quite the character, and loved to tell anyone who would listen about his escapades both in and out of jail.

At six in the morning on Saturday, I was released from jail, and that episode was behind me, except for the fact that my driver's license was suspended.

To remedy that situation I bought a Cadillac from a dealer on Livernois Avenue in Detroit and used the driver's license of Harold Strock, an American friend, so I could drive my new American car in Windsor.

One day when I was drinking, it was raining very hard, and I was stopped at a red light. Next to me was a policeman on a motorcycle. I rolled down my window and asked him, "Aren't you getting awfully wet sitting there in the rain?"

The officer said, "Never mind the rain. Just pull over and show me your driver's license."

I showed him Harold Strock's American license. The officer said, "That's not your license. I know you, and I want you to follow me to the police station."

At the station the officer took me inside and told the desk sergeant, "I'm charging this man with driving a car with an improper driver's license, and I'm going to impound the car and ticket him accordingly."

I called Ted Boomer from the police station and told him what was happening. Ted asked, "Who's the sergeant on the desk?"

I asked the sergeant his name, and then he talked to Ted on the phone. Ted told me he was coming right down to straighten things out.

Ted came down, and it turned out that he and the sergeant had been in the war together and were very good friends. Ted said, "Pete's a close friend and business associate of mine, and I would like you to let me get his car out of impound."

The sergeant said he would have it done right away. He gave me my license back with permission to get my car. The patrolman who had brought me to the station was standing there, swearing and saying, "It can't be done."

The sergeant told him, "Get back on your motorcycle and keep on doing your duties as a patrolman. Just forget all about this little episode."

I didn't get charged, and nothing more happened. The irony was that I had been drinking, and yet the motorcycle policeman had me drive myself to the police station.

Sometime later, I got pulled over by another policeman who knew me and thought I was again driving without a valid license. By luck or by chance, I had just gotten my real driver's license back that week.

GM Diesel Instructor's Course

In the 1950s, diesel engines were just coming into service on the railroads to replace the steam engines.

The operation of the new diesels was very strange to the engine crews, and they had trouble adjusting to these new engines. The new cabs were very comfortable with swivel armchair seats. They also had heaters, clear vision windows, and a windshield wiper. All the engineer had to do now was operate the throttle and the air brakes. When a problem came up, alarm bells would ring to tell the crew about the ground relays, etc. No longer did the engine crew need to wear overalls that got extremely dirty. Now they could ride the diesels in regular clothes that stayed clean.

Diesel instructors were needed at that time to ride along with the engine crew to educate them about their operations. In reality, though, it was similar to operating today's automobiles.

Peter James McLean

At this time, I was sent to the General Motors' training school in LaGrange, Illinois, to become a diesel instructor. This initial course took three weeks to complete.

LaGrange is located on the outskirts of Chicago, and a fellow classmate by the name of Percy Pinder had been born and raised in Chicago and knew the hot spots in the city very well. To get downtown Chicago, you just had to step on the Go train, and the ride would take less than thirty minutes.

During the day, we would plan which bars or taverns we were going to that night. Everyone would put a certain amount of money into a general pot, and that was the money Percy would use to take us on our travels to the hot spots in Chicago.

In those days, the bars and taverns had go-go girls dancing for the entertainment of the patrons. During our tours of the city, the girls would sit with us during intermission, and we got to know them, generally speaking.

One day I told the boys that I had to return to Detroit on the weekend and was booked on the number forty-eight train headed for New York. When Percy heard me comment on one of the girls, saying she was very beautiful, he and others hatched a practical joke on me.

The number forty-eight train was known as the Bawdyhouse because it was made up of only compartments and a diner car. It also had priority status and was guaranteed to be in New York no more than one hour late. Because of all the compartments on this train, a lot of businessmen took along their girlfriends and secretaries.

When I boarded the train, the go-go girl I had admired earlier was waiting for me in my compartment, along with a couple of cases of canned beer. The girl asked, "Is it okay for me to ride with you to Detroit since I'm going to visit some friends there? Percy and the boys chipped in to cover my fare."

Since it was not costing me anything, I said "okay".

During the night we were drinking, and when a beer can was empty we would just drop it on the floor of the compartment. What I didn't realize was that the compartment walls in the sleeper were open at the bottom about 4 inches. That meant that the cans we were dropping on the floor would roll from one end of the coach to the other when the train stopped and started. This happened all during the trip from Chicago to Detroit.

At some point during the night the girl asked me to come with her to her friend's place in Detroit, but I made no commitment.

When the porter called out that Detroit was the next stop, I got up and packed my bag and got ready to depart the train. As I was leaving, I passed the porter, who asked me "Holy Jesus, Peter, was that you who kept dropping the beer cans on the floor? I just spent the entire night picking them up!"

"I hate to admit it, but it was me, and I didn't realize the cans would roll back and forth the whole length of the coach."

"Well, I certainly had one hell of a time picking them all up."

I tipped the porter well in order to make him feel better and realized I hadn't done it on purpose. The porter was very happy when I left.

When I left my compartment, the girl was sitting and putting on her makeup. I went back and told her to hurry up because we had to get off the train. She insisted I'd just have to wait until she was finished what she was doing. I left then and stepped off the train onto the platform, and when I looked around, there was my boss.

I didn't know that during the night they had been asking if I was on the number forty-eight train because they needed me. A diesel freight train was stuck in the Detroit/Windsor train tunnel. They needed to get that freight train out of there so that the number forty-eight would be on schedule.

Tommy Cotrow was my boss' name, and he said, "Peter, we've been waiting for you. I need you to go down into the tunnel and get the diesels started, and then I want you to ride the freight train right to Black Rock in Buffalo to make sure there are no more breakdowns."

I said, "Okay, Tommy, let's go," and we left the platform, jumped into his car, and drove to the train tunnel.

I was very nervous standing on the platform with Tommy, and I was in a hurry to get going before the girl disembarked from the train. As luck would have it, she didn't get off the train before we left, and I never saw the girl again.

Incidentally, the boys back in LaGrange were all laughing and talking amongst themselves, because they knew that management had been calling for me and that they would be meeting the train in Detroit. They were waiting to hear what happened when it was discovered that I had a go-go girl from Chicago in my compartment. They thought for sure I would be fired or at least called on the carpet and would have a lot of explaining to do.

What they didn't know was that the girl had stayed on the train so long putting on her makeup that Tommy and I had already left before she showed. The boss never saw her.

I got the diesel out of the tunnel, and we proceeded toward Buffalo. There was no further trouble with the train once I got the ground relays taken care of and the train underway.

When I got back to the training school in LaGrange, Percy and the boys couldn't wait to hear the story of what happened in Detroit. When I told them that nobody saw the girl, it took some of their enjoyment out of the prank that they had set up for me.

While talking about the trip, I realized that the girl was not as beautiful as I first thought. Daylight was not kind to her.

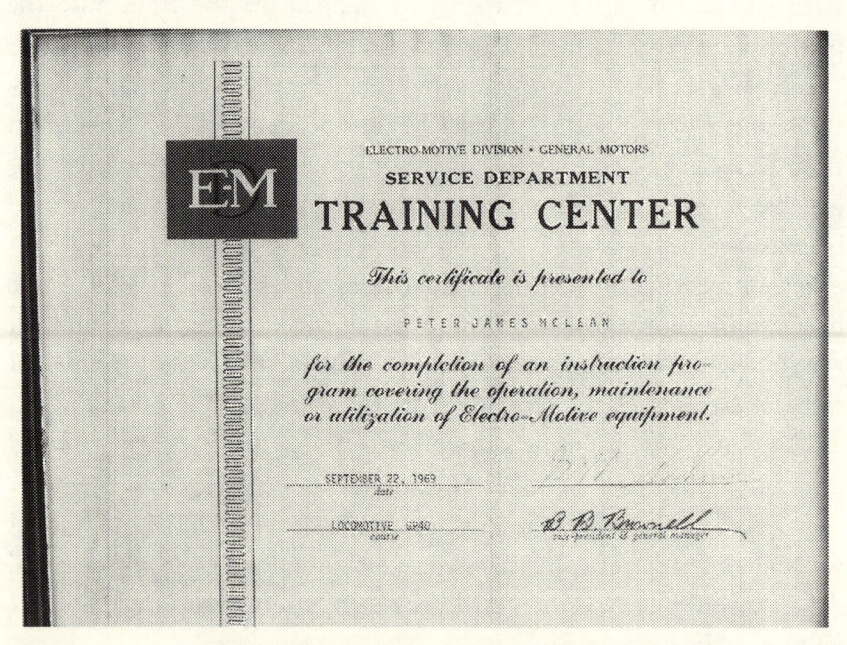

Certificate from Electro-Motor Division
General Motors Training Center

Henry Ford Hospital Dentistry Department

All during the time I was at school in LaGrange, Illinois, my teeth were acting up, and I knew I had to get them fixed, because they were causing me a lot of pain. My teeth were aching all during the trip to Buffalo and back to Detroit. I made up my mind that I was going to go to a dentist to have my teeth fixed as soon as I got back home, but first, I had to finish the course at the training school.

I had heard about the dentistry department at Henry Ford Hospital in Detroit, which was located in the vicinity of West Grand Blvd. and Woodward Avenue. In their dentistry department, there were specialists for everything. I figured that with the trouble I was having with my teeth, it was the best place to go to get them taken care of.

As soon as I finished the classes in LaGrange, I told the boss that I had to have a few days off in order to get my teeth fixed at Henry Ford Hospital in Detroit.

For my first appointment, I went over to Detroit early so I would have time to stop in a bar that was located close to the hospital. I needed to have a drink or two before I went to my dental appointment.

In the bar that morning, there were some guys who were all excited and talking about a long-shot race horse that they knew and that it was a sure bet. The bartender, an active bookie, accepted bets on horse races, and he seemed to do a good business.

I said that the horse sounded pretty good to me, and I gave the bartender ten dollars to put on the horse to win. Then I proceeded up to the tenth floor of the hospital to get my teeth examined.

After my teeth were examined, the dentist said I had lots of trouble because I had pyorrhea. They recommended that after the treatments to take care of the infection, I had to have all my teeth out. They said they would immediately insert the false teeth right there in the hospital. I made about five trips back to the hospital so they could prepare my mouth before pulling all my teeth.

All during this time, I had the habit of stopping in the bar and having a couple of drinks before I went up to the dentist for treatment. The day after I had bet on the long shot in the bar, I went in, and the bartender said that the long shot had come in and I had won two hundred and some odd dollars. I told him to keep the money and I would use the credit to take care of the expenses of my drinks.

I felt pretty good about the whole situation. I felt that I was lucky to be over there having my teeth treated and all my drinks paid for the duration of my treatments.

Finally, the dentist told me that I was ready to have my teeth extracted, and as I was told earlier, the dentist would insert my new false teeth right after the procedure. The doctor commented that every time I came for my appointment, he had smelled liquor on my breath. He cautioned me that when I came for the removal of the teeth, if I had any drinks beforehand, the freezing wouldn't be as effective as it should be. I told him I understood perfectly.

It turned out that in my account at the bar, I had approximately the cost of twelve drinks left to my credit. On the final day, when I was going to get my teeth removed, I naturally stopped in the bar and figured I might as well use up my credit, because it was not likely I would get back there again.

So up I went to the tenth floor. I was full of courage from the drinks I'd had, and I told the dentist I was ready to have my teeth removed. He reiterated that I'd probably have some pain because of my drinking, but I said to go ahead and do it, because I was ready.

They removed all my teeth and inserted my new false teeth, and things didn't feel too bad, because my mouth and gums were frozen. I immediately returned home to Windsor.

On the way home, I bought a couple of cases of beer to protect me for the next day, because I knew that when the freezing came out of my mouth, I was going to be sore for a few days.

When I got home, it was raining, and the driveway at the house had shallow pools of water on it. It was summertime, and I changed my clothes because of the warm weather. I went back out to my car in my bare feet and lifted the cases of beer out of the trunk and set them down on the laneway in front of the garage, where I was going to keep them.

By mistake, I had set one case in a shallow puddle of water while I opened the doors to my garage, which housed my den. I went to pick up the case, and because of it being all wet, the bottom fell out of the case, and the beer bottles broke all over the driveway. I got a large cut on my one foot, and the cut was deep enough that I knew I had to get stitches.

I called our neighbor, Dr. Ev Morris, and he agreed to come over to the house and examine my foot. He said I needed at least ten stitches and that the best place to do it was right there at the kitchen table.

He went back home and got his medical bag, and when he returned, he propped my foot up on the kitchen table. He cleaned and disinfected the cut and proceeded to stitch it up.

"Now, Pete," he said, "I want you to go right up to bed and keep your foot elevated, because your foot is going to give you a lot of pain while it's healing."

Also, the bad news he had for me was that I was going to have to be in bed for about a week. I didn't believe him, and I thought I'd be able to walk on my foot the next day.

By the next morning, my foot was swollen up to twice its size, and I couldn't walk on it. The most I could do was limp to the bathroom and then back to bed. *Well,* I thought to myself, *I guess I'm not so lucky after all.* There I was, stuck in bed with a sore foot I couldn't walk on, my mouth hurting something fierce and my hangover raging. The next week was looking pretty bleak for me.

The only thing that saved me was my friend Ted Boomer. Every other day Ted would bring a pint of liquor for me to sip on. It was just enough to keep me going.

Finally, my foot healed, the stitches were taken out, my mouth healed up, my teeth were working well, and my hangover was gone. I was back in business.

Daughter's Wedding

In October 1963, our youngest daughter, Carolyn, was planning a wedding, and she announced she was getting married in a Catholic church. Since I agreed to pay for a sit-down dinner for 250 people, I wanted the reception to be in a place of my choosing. That choice was the Masonic temple. I made the arrangements and discovered this would be the first wedding at the hall where alcohol was to be allowed.

When I applied for a liquor license for the reception. I received a letter from the Liquor Control Board of Ontario stating I was unable to obtain such a license for the period of twelve months from the date of the letter. The letter was dated August 24, 1962. This was done because of my drinking, traffic tickets, and various other infractions. At the bottom of the letter it read, "This action has been taken in the best interest of your family."

However, the wedding was not until October 26, 1963, which, it turned out, was two months past the period I was prohibited from obtaining liquor.

The wedding was a success, even though I might have had too much to drink during the day and evening festivities. I'm told that when I handed over my daughter to the groom, I told him quite loudly, "If you back out of the wedding right now, you can still come to the party!"

Dr. Alewick

Our family doctor at one time was Dr. Alewick, and he was very friendly and obliging in treating me in one particular case of overindulgence in alcohol. He said I was a dipsomaniac, which I learned was someone with an uncontrollable craving for alcohol. I thought at the time that this sounded better than just being classified as an alcoholic or a drunk.

I tried to cure myself, thinking I could drink one bottle of beer every half hour to keep my nerves and my hangover under control.

I wasn't quite ready to give up drinking, so I continued to get myself quite inebriated. Because my condition was becoming so critical, an ambulance was called. I refused to be taken out of the house, so my daughter Barbara, who was visiting, tucked in a bottle of whiskey beside me on the stretcher, and away we went. Of course, the bottle was taken away from me as soon as we got to the emergency room.

When I was checked into the hospital, Dr. Alewick told me that during my treatment he was going to try something new. He was going to give me a double shot of liquor at breakfast, lunch and dinner. I said it sounded like a good idea, and I thought it should help.

The first time the nurse taking care of me brought me the liquor, she brought it in a shot glass that she was waving around on a tray as she talked to the other nurses. The shot glass was supposed to be full of liquor, but by the time she got it to me, she had spilled about half of it. I called her a "stupid bitch" and told her the doctor wanted me to have two full shots.

She said, "Well, Peter, I don't like you either."

"The next time you bring me the liquor, bring it in a big glass."

She brought me a larger glass once and then reverted back to the shot glasses that she kept spilling. I told her over and over again to bring it in a big glass, but she kept saying she forgot.

She also said "I don't believe in this type of treatment at all." Before we were done with the liquor, the nurse and I hated each other intensely.

Finally, I told Dr. Alewick, "You might as well cut out the liquor treatment, because the nurse spills half of it before I get my hands on it."

I followed it by saying, "That nurse should be fired from the hospital."

Dr. Alewick ignored what I was telling him and said, "I think you are recuperating quicker than I thought you would, and I'll be sending you home soon."

A week later I was fully recovered and back in good shape. As I was checking out of the hospital, the nurse that I had hated so much made a special trip down to say good-bye to me.

She said, "Peter, I disliked you immediately because you should have been ashamed of yourself because of the shape you were in. Every time I looked at you I was reminded of my husband who I hate because he too is an alcoholic."

As I was going out the door, she said, "You look good now, and you are off to a good start. I certainly hope you've learned your lesson."

I too hoped that I had learned my lesson. But, suffice it to say, I didn't stay on the straight and narrow for long.

BOATING FOR DUMMIES

The "Marion B"

I felt it was time to do something just for myself, so in 1959 I bought a twenty-eight-foot speedboat that was built and owned by H. P. Keller, the president of the Chrysler Corporation. The boat came equipped with twin 302-cubic-inch six-cylinder Continental engines used in the Hudson Hornet cars. The layout of the engines made testing and working on the engines easier.

To increase speed and power, I replaced the engines. At a marine supply factory in Coldwater, Michigan, I had two of the big Chrysler V-8 Hemi engines with superchargers put in. With the new engines, the boat was overpowered and a real danger. It was faster than any coast guard craft.

Today, everyone knows the dangers of drinking and driving (in a car or in a boat). At the time I had the boat, I was drinking quite heavily and used the boat recklessly and it caused me a lot of grief while I had it.

I named the boat the *Marion B*, and I kept it docked in a covered boat well on the east side of Windsor.

Joe Masco was a friend of mine as well as a reporter for the *Windsor Star*. He would often stop by for a drink on the boat. Joe's favorite was beer, and he was a real bum when it came to his beer.

One Sunday morning, there was no beer to be found at the boat dock, and I made a bet with Joe that I could go to Detroit, get beer and liquor, and be back within twenty minutes. We made the bet for a tank of gas in the boat.

I took off with the *Marion B* and headed to McGregor's boat dock across the river in Detroit, Michigan. I often got beer and liquor from McGregor's, and if he knew I was coming, he would have everything sitting on the dock waiting for me. I just had to load up the boat, hand them the money, and be gone.

The day I made the bet with Joe, I phoned McGregor's and told them I was on my way and to have the stuff waiting for me. So over I went, picked up the supplies, paid for them, and took off back to Windsor. I made the trip in eighteen minutes. I remember that because Joe owed me a tank full of gas for my boat, which held 30 gallons. To get him to pay for the tank of gas, I had to shame him into upholding his part of the bet.

Joe was a very likeable person, but he was killed in a car accident driving from Tilbury to Windsor shortly after he lost our bet. I arranged a party and raised money for Joe's wife. We wanted to make sure she was financially stable enough to get by until she got all her affairs in order after his death. I had never met the wife, and it wasn't until after we gave her the money we raised that we learned she had already made arrangements to move to western Canada with her new boyfriend.

What a blow! Dick Harrison, my other reporter friend, didn't seem too surprised by this turn of events, and he told me to just let it go and move on.

Another interesting boating friend of mine, Norm Black, owned a boiler factory in the Detroit area, and he had a thirty- to thirty-five-foot boat named *Cutty Sark*. Norm was always talking about making his boat run faster. However, he wanted the speed but not the cost. He would run the boat to the Windsor Yacht Club and park it there for the day.

Norm had a different young woman in a bikini on board to serve drinks each time he docked in Windsor. He always had an unlimited supply of Cutty Sark scotch whisky on board. It was all he served— never beer unless you brought your own.

If he had a wife and family, he never talked about it. We had some fun times, however, just sitting around on his boat. He never wanted to take it for a run except to get it back to his dock in Detroit. That seemed to me to be the extent of his boating.

Peter James McLean

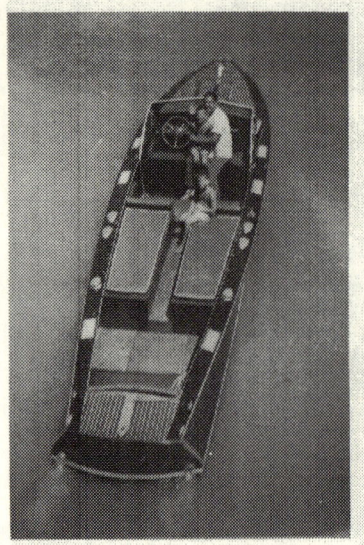

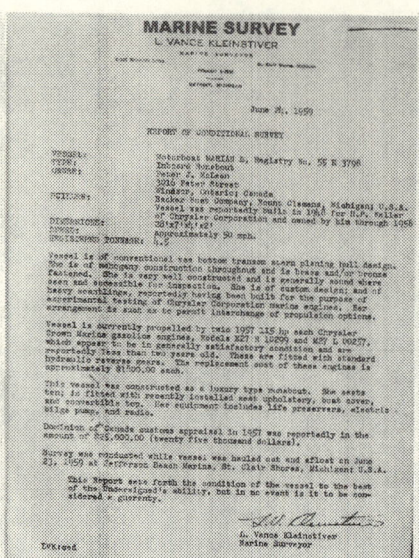

Right: Joy riding with my two boys on the "Marion B"
Left: Marine Survey of boat when purchased

Running Without Lights

Sinbad's is a popular US watering hole catering to the boating crowd. One night I was coming back from there and heading for the covered shed where I stored my boat when suddenly I rammed into a yacht. I didn't have any running lights on and hadn't seen anything on the water, so this was a real jolt.

My windshield broke, and I had a bad cut on my arm that was bleeding heavily and needed to be stitched up. At the same time I saw the yacht immediately turn over and start to sink.

I had a friend, Nick Smith, with me and he was uninjured. I had rammed the yacht right in the side with the bow of my boat. I only had a small hole above the water line, so my boat was in no danger of sinking.

There were two men, a woman, and—I can't forget—a dog on the yacht, and they were all in the water. Nick was an excellent swimmer, so he immediately dove in the water and grabbed the lady and handed her to me. I pulled her aboard my boat. The other man from the yacht could swim, and he came to the side of my boat, and I pulled him aboard as well.

Confessions of an Eccentric Dreamer

All of a sudden the woman climbed on my back and was scratching me, tearing my sport shirt.

She kept screaming, "Save the dog! Save the dog! Never mind my husband—save the dog."

I hollered to Nick, "Grab that man. He looks like he's drowning. Never mind the goddamn dog."

Nick grabbed the man and got him to the edge of the boat, and I pulled him aboard and pumped the water out of him. He turned out to be okay. The last thing we did was to scoop the damn dog out of the water.

The woman kept up her screaming and continued to physically attack me, creating quite a commotion. The other man from the yacht whom I got out of the water first did nothing, absolutely nothing. He just sat there and watched like he was a stranger to everybody. Here was a man who could swim and did nothing to help the owner of the yacht. He also did nothing to help me when this hysterical woman was so out of control.

When I got us to my shed on the Windsor side of the Detroit River, I said I had to go to the hospital to get stitches because the large gash in my arm was still bleeding heavily. No one else needed medical attention, so I went to my car and got behind the wheel.

Again, this woman attacked me, ranting and raving and making another huge commotion. She tried everything she physically could do to stop me from leaving. Even while I was in the driver's seat, she attacked me again and tried to pull me out of the car. Finally, I had to brace myself on the armrest of my Cadillac, put my left foot on her chest, and then shove her away from me. She ended up skidding about one hundred feet across the driveway.

She didn't care that I was still bleeding and needed to get to the hospital. Again, the other man from the yacht just sat there enjoying the show. The owner was unhurt but exhausted from his experience and unable to help, but he did say he didn't need to go to the hospital to be checked over, so I finally left on my own.

The police came to the hospital to interview me, but they didn't request a blood test for alcohol. However, they did say I could be charged with careless driving or something similar. After the interview and getting the stitches in my arm, I went on home.

I didn't learn until the next day what had happened to everyone. Nick told me the three people got in a cab and were having the driver take them to Wyandotte, Michigan, where they lived. I was not concerned, because they could well afford any expenses incurred and were uninjured.

When I appeared in a Windsor court regarding the case, I thought they would throw the book at me, and they did find me guilty of careless driving. They opened me up to being sued by the yacht owner and liable for all expenses incurred retrieving and fixing his yacht.

While I waited for court to commence, a stranger come over and started talking to me, asking where he could find me after court. I told him we were going to Fisherman's Cove at the Prince Edward Hotel, and he said he would see us there. None of us knew who the man was at that time. Later he did show up and informed me he was the owner of the yacht I had sunk, and he then ordered a round of drinks for everyone.

The man took me to another table and informed me that he would not be suing me. He said he was so glad he found out how his wife really felt about him that he was in the process of getting a divorce.

Who would believe this one! It turned out the man owned a shipyard, and the yacht I sank was worth well over $250,000 at that time.

He said he had kicked his wife out of the house that very night and that had it not been for the accident, he never would have learned his wife loved the dog more than him.

"And," he said, "You did actually save my life, because I can't swim and I was on the verge of drowning. Here you were trying to help me, and you had to deal with my wife, who was trying to prevent you from saving me and a friend who did nothing to help anyone."

True to his word, he never contacted me again.

Raging Lake

My friend Harold Newton and I were going to Middle Sister Island in the *Marion B* because I wanted to take some pictures of the island. Newt was a friend and machinist on the railroad and worked for me in the engine house.

We stopped at Duffy's Tavern in Amherstburg at about three in the afternoon and asked the "weatherman" who was usually there drinking how the weather looked. We did this as a precautionary measure. He said it was going to be clear and sunny, so after having a few beers, we left for Middle Sister Island.

Middle Sister Island was a half hour running time from the mouth of the Detroit River. When we left Duffy's, the weather was just as the "weatherman" predicted. We were cruising on smooth water, so we got to the island in record time.

Newt wanted to take his time walking around and admiring the beautiful scenery and the silence. When you were on the island, the silence was the thing you noticed most. There were no outside noises to disturb the beauty of it.

At that time, I had an option to buy Middle Sister Island. I was planning to sell the island to a club of some sort that would like the privacy of the island for their buildings, etc. I never followed through on actually purchasing the island and later regretted it.

At one point, I saw black thunderclouds over the horizon coming from Toledo, Ohio, so I said to Newt, "It looks like there is a storm coming, so we better high tail it back to the shelter of the Detroit River."

It was about six in the evening when we left the island, and we were running in daylight.

We were about fifteen minutes running time from the safety of the Detroit River when the sky turned dark and menacing, and the change was quickly followed by thunder and lightning. We knew immediately we were in for a bad storm.

We were cruising at about forty miles per hour when the storm hit us with very heavy wind and rain. All of a sudden there was a rogue wave that hit the bow of the boat, knocking Newt back into the stern and knocking me down between the two engine hatches. I climbed back into the driver's seat and found there was only one engine working and that the waves were getting unusually high.

Newt, in the stern, was up to his waist in water. I hollered to him to start bailing, because I thought the boat was going to go under any minute. I began to run the boat in the direction of the waves. It was the only direction I could go, because we only had one engine. The waves were so high that I couldn't run through them. I had to keep the bow from breaking into the wave itself.

It was now pitch-black out, and the only thing we could see was a sparkle of lights coming from Colchester, Ontario. The storm had carried us that far in a short amount of time. I couldn't turn the boat in that direction, because I'd be crosscutting the waves, which were really rolling. We took off our shoes and shirts, and I threw Newt a life jacket and, for the first time in my life, put one on myself. I told Newt we were sure as hell going into the water.

We kept going, and as long as I had the one motor running, I could keep the boat in the trough of the waves so they weren't breaking over the bow. About half an hour later, we were still going, and when I looked back to see how Newt was holding up, I just couldn't believe what I was seeing. I was looking straight down! Here I was in a twenty-eight-foot boat, and I had to look straight up to see the bow of the boat and then straight down to see the stern.

I'd never thought Lake Erie could generate such huge waves. There was more water covering the stern, and Newt continued to bail like a mad man. I could tell he was becoming exhausted.

I finally saw some lights and they were in the direction we were going, but I couldn't tell where they were coming from. The water just kept rising in the stern, and we had to get to those lights if we were going to survive.

About an hour later, we came up close to the lights, and I could see that it was a tugboat pulling a barge with a drilling rig on it. It was trying to pull the barge into the safety of the harbor at Kingsville, Ontario.

I had a hundred-foot-long line on board and yelled to Newt to throw it to me. I tied one end onto the tug without seeing anybody about. After tying us off, I told Newt we were staying right where we were until the storm was over. We were very, very lucky. We had thought we were goners.

We climbed down into the cabin of the tug and found the captain. We explained how we had gotten caught up in the storm and how we ended up being tied to his tug. I told him we couldn't go anywhere soon because my boat was unsafe and would sink if we tried. The captain could barely see the outline of my boat, and he welcomed us aboard.

"Well," the captain said, "consider yourselves very lucky that you got this far, because before this storm is over, we're going to hear about some boats being lost on Lake Erie."

The captain then told us that he wasn't going anywhere because his tug couldn't haul the drilling rig in this storm. He was waiting for another tug to come out from Kingsville to tow the rig into the harbor. We were extremely glad to be sitting in the shelter of the tug's cabin. We were just like two drowned rats.

Finally, a couple of hours later, the other tugboat arrived. They set it in towing position and towed the rig and my boat into the harbor.

When we were finally safe on land, I went back and looked at the *Marion B* as it sat there alongside the dock. It was about half full of water but still floating.

Newt and I called a taxi and went to one of the hotels in Kingsville. Newt called his wife, Aggie, and she said she was so glad to hear from him because the coast guard was on the lookout for us, and everyone thought we were adrift in the storm. Aggie then contacted the coast guard and told them we were safe.

Although I didn't have my wallet, I had money stuffed in the watch pocket of my pants, and we were able to stay in a hotel for the night.

The first thing we did in the morning was check out my boat. I said to Newt, "I'm going to bail the water out and see if I can start the other engine so we can run the boat back to Windsor."

Both engines worked just fine, and the return trip was uneventful.

I had never seen Lake Erie so rough, and I would never have believed a storm could come up on the lake so quickly. It had gone from daylight to pitch-black within ten minutes.

Fishnets

One summer, Marion, her parents, and I rented a cottage in Port Stanley, Ontario, a community on Lake Erie that was a very popular summer destination—not as popular as it was when Marion and I were young, but popular just the same.

I thought it would be a good idea if I ran the boat down there so we could use it while we were there. I thought I could make the trip in the daylight and that if I just ran a couple of miles off the shoreline that everything would be okay.

The weather had turned stormy when I left about four in the afternoon, but I was planning to be in Port Stanley within four or five hours if I was running in smooth water.

However, what I didn't realize was that to clear Pointe Pelee, I had to go twenty-eight miles out into Lake Erie because of the extensive sandbars at the end of the point. After clearing the point, there was a strong wind blowing from the south that made the Lake very rough, and by that time, it was quite dark and hard to make good time.

I had run through two sets of fishnets that were out in the lake, and because it was dark, I couldn't see the marker buoys for them. It slowed the boat down considerably. I knew that I was close to the harbor at Wheatley, so I headed in toward shore. I missed the harbor in the dark and ran up a small creek that was used for outboard motor boats. It was very shallow, and the boat got stuck in the mud.

The people in one of the nearby cottages heard the roar of my engines and came out to investigate.

One fellow said, "You just better leave your boat until daylight and come in and stay in the cottage. In the morning, I'll take my boat, which has an outboard motor, and we'll tow your boat back to deep water, and you can continue on your trip."

In the morning, he got his motorboat in the water and helped me get my boat out of the mud. Having never met me before, they couldn't have been nicer or more welcoming. I was grateful for their help.

The one thing I never considered when making this trip was that the fishermen had their fishnets strung out in different locations on

Lake Erie. The fishnets were so strong that they wouldn't break when a boat ran over them and would wrap around and get caught in the props.

The next day I made it to Port Stanley, where I had to have the boat lifted out of the water at the marina. Both propellers were covered with fishnets, and the nets were strong enough that they had bent one of the drive shafts.

I got Doug Buck, who had a machine shop, come down and help me pull the bent driveshaft out of the water. He cleaned off the fishnets, and the boat was just like new again.

For some reason, I never had safety glass put in the windshield, not even after the episode with the yacht where I cut my arm so badly. Again, the windshield had to be replaced, and like before, I didn't put in safety glass.

I realized I needed to find out more about these fishnets and how to recognize them out on the lake before I took the boat back to Windsor. I found the captain of a fishing boat, who was extremely helpful and answered all my questions. He taught me how to read the buoys that marked the nets and in which direction the nets were strung. Boaters should know how to read the buoys so that in the daylight they can watch for them.

The boat was running real good and we had a good time in Port Stanley. When it was time for me to take the boat back to Windsor, I decided to leave in the early morning so I would be certain I was running in daylight.

The return to Windsor was pleasant and uneventful.

The Final Run of the Marion B

The Rooster Tail was another well-known restaurant/bar in Detroit, Michigan, that catered to the boating crowd. It was August 22, 1960, at two thirty in the morning when a couple of friends and I left the Rooster Tail after a night of partying and were heading back to Windsor in my boat.

I came up behind a cruiser that was creating a ten-foot wake behind it. The bow of my boat went into the wake, and it pulled the bow over toward a cement dock at the end of Belle Isle. My bow was at the same angle as the cement break wall, and the boat slid right up it. Because of the speed of the boat, it slid right through a fence and went down the other side in the area of a US Nike missile base. Nobody in the boat was injured, but my boat had a broken keel.

Not many people even knew the missile base was there, but when my boat went through the fence, all the alarms were triggered. I didn't want to be caught by the guards, so I got into the water and started to swim away from the dock. The guards just stood on shore watching me, because I had nowhere to go. Finally when I was too tired to go any farther and had to swim to shore, the police picked me up.

With the security alarms triggered at the base, the whole story ended up on the front page of the *Detroit Times*. This was followed by an article in the *Windsor Star* reporting that a "Canadian Assaults Belle Isle Break Wall."

I got a call from Joe Masko, my reporter friend at the *Windsor Star*, and he asked, "Did you see the Detroit newspaper?"

"No."

"There's a picture of Judge Watts on the front page saying you are going to jail over this boating accident."

"Thanks for telling me Joe, I'll talk to you later."

I already had a reckless boating charge against me for buzzing boat wells at the foot of Fairview in Detroit, Michigan, causing damage to boats moored there. This was in July, and I had never showed up to answer those charges in the traffic and ordinance court.

I called George Yates, my lawyer, friend, and neighbor, and I told him, "I just had an accident with my boat, and there's this article in the Detroit paper where a judge said he has to clamp down on these boat operators and that he understood that I had previous tickets for reckless driving. I have to go to court in Detroit tomorrow, and I think they might put me right in jail. I need you to come with me."

George said, "Okay, I'll pick you up at 8:00 a.m."

The next morning, George and I went to the courthouse in downtown Detroit. We were early for the case, so we stopped in a bar to have a couple of drinks beforehand to steady our nerves.

When we got into the courtroom, George told the judge, "I'm a Canadian lawyer and am not licensed in the US, but it would be appreciated if you could give us an adjournment of the case until we can engage a US lawyer. Mr. McLean had been traveling and could not appear in court regarding the earlier charges."

"Yes, you can have all the time you want, Mr. Yates, and I would like to welcome you to my court and tell you I appreciate your coming here."

Judge Watts was very obliging, and the case was adjourned for two months.

I hired a US attorney named Nobel Lawson in the Buhl Building in downtown Detroit. He had done previous business matters for me, and when I told him the circumstances of the case, Nobel said, "I'll handle it, Peter. In fact, I know Judge Watts personally."

"When will I be going to court, and what should I do now?"

"Just go home and wait until you hear from me."

In the meantime, a Windsor boatyard owner and two employees were arrested by the Detroit harbormaster for trespassing when they attempted to salvage my boat. They claimed I had given them authorization to remove the stranded craft. I can't remember giving them authorization, but with the shape I was in anything was possible.

About three weeks later, Nobel called me over to his office, and when I got there, he said "Your case, Peter, is settled, as long as you provide the following two items: a statement from everyone who was on your boat at the time of the accident certifying they were not injured, and a receipt from the parks department on Belle Isle showing that you have paid for all the expenses of repairing the fence, lifting your boat out of the water, and putting it in storage on Belle Isle."

When I turned these documents into the court, the case was completed. There was no fine, and I didn't have to appear in Judge Watts' court again.

It turned out that the judge being on the front page of the *Detroit Times* that day when the accident happened, was mainly for election propaganda. The outcome was a complete surprise to me.

I was very, very lucky, because I did have speeding and reckless driving tickets from the US Border Patrol. They always liked to stop and check out the boat whenever I was out on the river. Then, when I had the larger engines installed in the boat, it did not make them very happy, and they were always on the lookout for me.

I ended up selling the boat piece by piece because it contained very expensive Borg Warner hydraulic transmissions, and I made good money doing so.

I think these stories reiterate that having a boat that is overpowered is much too dangerous, especially for an alcoholic.

The Marion B Lives Again

The hull of the *Marion B* ended up in a barn in Leamington, Ontario, and I didn't hear anything about it for many, many years.

In 2000, a man by the name of Jack Davidson bought the hull and planned to completely restore the boat. He spent about five years having it worked on. Family friends Stan and May Young, who knew Jack, recognized the boat and told me what was being done with it.

They put me in touch with Jack, and I was able to spend some time with him discussing the boat and the work he was doing on it. I was able to give Jack the history of the boat, along with copies of the original documentation. He was very pleased.

Over a five-year period, he had to have spent a great deal of money to get the boat completely restored. There are not many people with the kind of money and determination it took to do this restoration.

Jack was battling cancer and had made it his goal to have the boat running on the Detroit River before he died.

Just prior to his death, the *Windsor Star* had a picture of him on his boat when they were doing the test run. He got his wish.

ACCIDENTAL DISCOVERY

How did I manage to go from railroading to prospecting in Northern Ontario? It amazed even me how it started.

My parents owned a lodge on an island at Ashigami Lake in Scadding Township near Sudbury, Ontario. They used the cabin there for fishing and hunting every year. I thought I would like to have my own cabin.

I started researching available land, and in the *Ontario Gazette*, I found that there was a large auction of land coming up in Hastings County in the Bancroft area.

I had never been to Bancroft, but I did think that it would be a good area for hunting and fishing. I knew that during the hunting season in the Bancroft area, moose, deer, and bear were all open at the same time. It was also a good hunting area because of it being close to Algonquin Park.

The auction was being held in Belleville, Ontario, the headquarters for all Hastings County, and I decided to go. Different properties were to be auctioned off to the highest bidder, just like an ordinary auction.

Since I had never been there before, I was comparing the lands for sale off a county map that showed all the townships and the lots and acreage, etc., of the different properties to be auctioned. I bid on properties that were the closest to Bancroft.

I was totally ignorant of what was on the properties and their value, so my bids were stated as taxes and costs, which meant the price I was bidding was the amount of taxes owing on the properties and the costs of the auction.

I bid on numerous properties, but my bids were too low. Then there were four lots in Dungannon Township consisting of approximately eight hundred acres of property that I bid on. I won that bid. Another of my bids—on a fourteen-acre property on the shoreline of Baptiste Lake, a large lake a little north of Bancroft—was also successful. I now owned four lots in Dungannon Township and fourteen acres on Baptiste Lake, but I had never seen the properties and wasn't even sure where they were located.

After paying the secretary for the properties, I asked him about the titles to the properties and what they included. It turned out that the surface rights, the mineral rights, and the titles were free and clear of any past mortgages or encumbrances as of the date that I had bought them. It was the best title that you could have on any property. It meant that to sell the properties, lawyers, etc., did not have to search the titles, because the titles were free and clear at the time of purchase.

The day after the auction, I drove up to Bancroft to search for and locate the properties I now owned. I checked into the Bancroft Hotel and asked Jack McAlpine, the owner of the hotel, who would be a good person for me to use as a guide to check out these properties. He told me about Cecil Dwyer, who was born and raised in the area, and said he should be able to show me the proximities of the properties.

The next day, I contacted Cecil, and he agreed to help me. We went first to Baptiste Lake and found the general location of the fourteen acres I now owned. The property turned out to be a whole bay area with no buildings, and I thought it would be a perfect spot to build a cabin on.

Then we went down to Dungannon Township, and I showed Cecil the map showing the adjoining lots. We kept looking, but in the end, Cecil and I never did find the four lots. Cecil finally said the location of the properties couldn't be determined without having a survey done, which would be very expensive. From what I could tell, the land was all bush, rock, and swamp. After three days of this, I went back to the hotel and was very discouraged to think I had bought property and couldn't even find it.

One night I was sitting in the beverage room of the hotel, drinking a bottle of beer and wondering whether I should hire a surveyor. I was

Confessions of an Eccentric Dreamer

completely undecided as to what I should do. Jack called me on the loud speaker and said there was someone who wanted to talk to Peter McLean. I told him to send the man in.

This stranger came in and asked me if I was Peter McLean, owner of four lots in Dungannon Township.

I said, "I'm Peter McLean, and I just bought them at auction in Belleville."

"Well," he said, "I represent Gray Hawk Uranium Mines, and I'm instructed to buy this property from you."

I asked, "Why would you be so interested in this property?"

He said, "I suppose you know that this property ties right into the continuing deposit of uranium that our company Gray Hawk Uranium is mining right now. We are producing very rich ore, and we are expanding our holdings as much as possible."

Then he said, "How much money do you want for the property?"

"Well," I said, "there are so many people looking at it that the property is worth at least $150,000."

When I said this, he thought that I was referring to other mining companies that were looking at the property.

He said, "I can't pay $150,000 for the property, but I could pay $100,000."

I said, "The titles on these properties, mineral rights, and surface rights are all free and clear. If we split the difference for $125,000 cash payment as the total amount for the property, you'll own it lock, stock, and barrel."

He asked me to stay there for a few minutes and wait for him to make a phone call. He was gone about half an hour, and when he came back in, he said, "Okay, the deal is made for $125,000."

I said, "All right, when are you going to pay me? Don't forget we're talking about cash money."

He said, "We'll close the deal tomorrow morning as soon as the Royal Bank opens—which, by the way, is located right across the street—and we'll use the bank manager as the witness for all the signatures involved. You'll be paid at that time with a certified check made out directly to you from Gray Hawk Uranium Mines."

I said, "Okay, the deal is made."

We shook hands and agreed to meet at ten the next morning at the Royal Bank.

During the night, I was turning the proposition over in my mind, and I was thinking to myself that I certainly hoped that this stranger, whose first name was Jim, was not bullshitting me. Here I was, having just purchased property I couldn't even locate, and here came someone who wanted to buy it from me for a large amount of money.

The next morning, I woke up, took the titles to the properties, and arrived at the bank at 10:00 a.m. sharp when it opened. Jim was there, all smiles, waiting for me.

He said, "Did you bring the ownership papers?"

I said, "You bet. I have them right here."

I signed the ownership of the property over to Gray Hawk Uranium Mines, and the bank manager had a certified check lying on his desk, all made out and ready to give to me.

I couldn't believe what had just happened. It turned out that Gray Hawk was one of the biggest and most profitable uranium mines in the area. I immediately had the bank manager transfer this money directly to my personal bank account.

Cecile Dwyer and me

The Road to Prospecting

There I was with a substantial amount of money, and the uranium boom in the Bancroft/Halliburton area was in full swing. If this was typical of the money involved, it hit me then and there that I should become a prospector myself.

With my newfound money, I bought my Lucky Strike Geiger Counter (my son Guy has it now) and then got some books on prospecting for uranium from the Department of Mines Library in Toronto. It was a terrific opportunity to get started studying the area and learning about uranium and how to prospect for it using my new geiger counter.

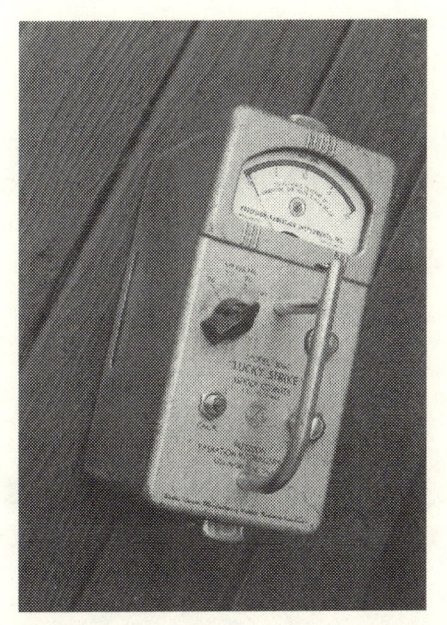

My "Good Fortune" Geiger Counter

Learning to Prospect

I quickly learned that prospecting is not done like it is shown on television—it is not done by searching in the bush. The first thing you have to do is to get up to date on the government reports.

I started out concentrating on the Bancroft areas and found out that there were reports from all the different townships involved.

During the winter, I would study the maps and reports, and before I left home to go north, I would know exactly where I was going. I could walk and stand right on the location as cited in the reports. From there, if the property looked good and was encouraging, I would go directly to the mining recorder's office to see if the mining rights on the property were open. If the mineral rights were open, I would go with a crew of two workers, and we would stake the claims.

I kept the fourteen acres on Baptiste Lake for a couple of years and then sold it. There is a picture of me with my mother standing on the lakeshore. It was a beautiful location, but I decided I didn't want to build a cabin there.

As a side note, over the years, people that owned land in the Bancroft area would give it to a neighbor or give someone permission for them to cut pulp wood, and they in turn would agree to pay the taxes on the property. You wouldn't be able to buy any of these properties, because you would never be able to track the true owners of them with so many people paying portions of the taxes.

Myself, my Mother and two of her friends standing on
The shore of Baptiste Lake

Cadillacs

I bring up my story here about my love for Cadillacs because of the reaction I got from many, many people. No one could believe what I did with them.

In 1952 I bought my first Cadillac from the proceeds from selling my first uranium property. I wanted a very comfortable car to drive on my long trips from Windsor to the Bancroft area.

When I first started prospecting, I would take my Cadillac into the backwoods and over rough terrain. They never let me down and went wherever I needed them to go.

Since that time, I've also owned a 1953, 1955, 1957, 1962, 1969, 1971, 1981, and 1989 Cadillacs, and right now I'm driving a 2000 Cadillac STS, which I haven't worn out yet. All of them had extremely high mileage, and I had very good luck mechanically with them. However, a couple of them were replaced because my mechanic refused to do any more work on them. He deemed them unsafe.

I have to admit that I did do some of the repair work myself. Anyone familiar with masking or electrical tape understands that it hides a multitude of sins. To hide rust spots on a really high-mileage car, I would cover them with layers of tape and then paint them a similar color as the car. To look at them from a distance, you would never know what I had done, but it was quite obvious up close.

In 1962, I saw a brand-new Cadillac sitting on a pedestal on a dealership lot in London, Ontario, and I decided to buy it. The dealer told me it had to be serviced and would be ready to go in two or three days.

When the dealer called to say the car was ready, I had Marion go down to London by train to pick up my new car and drive it back to Windsor.

I'll be damned, though, through a mix-up, she drove my brand new Cadillac on a freshly tarred road, and to top it off, she ran over a large dog.

When she got home, I told her, "Marion, that was a hell of a thing to do with my brand-new car."

I was sure I would have bad luck with that particular car, but it turned out to be one of the best Cadillacs that I've had.

Between 2005 and 2008, I kept looking for another Cadillac but couldn't decide what I wanted. I finally realized that I wouldn't be able to drive much longer, so I had to give up my search.

Throughout the book, I've referred to my Cadillacs and the fact that I've taken them into some really rough bush areas. Not one of them ever left me stranded. But I don't recommend others try this.

Pictures of two of my Cadillacs in the bush

Living and Working Conditions in the Bush

I quickly learned that I had to be able to get to the properties I'd claimed. I needed to have the proper vehicles, because there are few, if any, roads or trails. In some places, I even had to make my own roads using a bulldozer.

Over the years, I have had a four-wheel-drive Jeep, a Land Rover, a Volkswagen van, a half-ton Chevy truck, a six-wheel vehicle that could easily go through water caused by beaver dams, and a four-wheel ATV that could pull a small trailer. In addition, I had a covered snowmobile with tracks in the back and skis in the front (but no wheels). I bought this machine from the government as surplus equipment. It had been used in the Arctic region.

Eating and sleeping in the bush was often necessary. We used any mining shacks that were already on the property, or we made our own. It was a very rough life. We had to haul all the food and supplies we were going to need, along with our mining equipment, with us.

Types of vehicles I needed and conditions found in bush

BANCROFT ADVENTURES

I was now ready to pursue uranium mining on my own.

I quickly realized that to be prospecting for uranium, you needed township maps and government geological reports on the different townships and the area that you were interested in. I learned that one sign of a "hot" area for uranium was the presence of pegmatite, and you would use your geiger counter to locate high readings of radioactivity. Cecil Dwyer, with whom I became friends, was my guide going into the bush and locating different areas that were of interest.

The more I studied prospecting for uranium, the more I realized that the public perception of a prospector having to tramp through the bush was wrong. I was soon able, through the geological reports and geological maps, to pinpoint areas of interest. I would visit these areas, and using my geiger counter, I could determine if that particular area showed high radiation readings. I also bought a Capco surface drill, dynamite, fuses, and caps.

The first property I worked on was in Mount Eagle Township. We would drill holes approximately two feet deep in the dikes, and Cecil and I would blow them open and compare our radioactivity readings accordingly.

Within a couple of months, I found an exceptional property in Dunganan Township that was located about two miles farther into the bush, and I then drove into the property with an old jeep that I had at that time. I then realized that in order to have financing for a property this size, I would need a company with shareholders to be able to arrange the financing.

Using classified ads I ran in the *Detroit News*, I had an exceptional response from interested investors. At that time, I had a large garage that I had made up as a den with uranium samples from the property that I could use as a sales office for the American people I was talking to. Selling the shares in Canada was perfectly legal.

I set up a private company that would allow me to have fifty shareholders before the company would have to be turned into a public corporation. I would insist that the Americans I was dealing with come to where I lived in Windsor, Ontario, when I was selling them on the idea of buying shares in my company. At this time, I didn't have any partners in the company with me, so I had positions on the board of directors open and available to select people who were going to invest substantially in my company.

The company and the property turned out to be very encouraging. I had met a mining engineer from Nicklerim Mining Company in Sudbury, Ontario, who had an excellent reputation and name in the business. His name was A. J. McKinsey, and I had him do an engineer's report on the property.

One of the main shareholders in this company was a person by the name of Jack Oswald who owned a machine and welding supply business on Van Dyke in Detroit, Michigan. Jack was very aggressive, and he had solicited numerous friends of his to become investors in the company.

Jack had been to the property at Bancroft and had spoken personally with the mining engineer, and he became very enthused about the whole company. Jack had learned from me all the different things I knew about prospecting, staking claims, and studying maps and geological reports in regard to uranium.

The second time I took Jack up to Bancroft, he asked me, "How much would it take for me to buy all your shares?"

He wanted to buy me out completely.

The shares were listed with the government at one dollar par value. I told Jack I was interested in other projects in the area, and we made a deal that he would buy my shares at two dollars each and he would take over the company and use his own board of directors. He agreed with this because he was very anxious to get into the uranium business, and the deal was finalized.

That put an end to my first uranium company.

My Den and Workshop in Windsor

Uranium Mine Number Two

I was now free again to explore other properties to develop.

I found a good exposure of a highly radioactive pegmatite in Hershel Township. Cecil Dwyer and I did our drilling, surface blasting, and trenching. This property looked like an excellent prospect, and I decided to develop it further.

I started another company, which I called the Lucky Guy Exploration Company. I proceeded to raise some finances by selling shares in this company to Americans. Again, I had them come to Canada, and I would do the selling out of my garage/den in Windsor, Ontario. The uranium boom was getting stronger, and everyone was interested in investing during this time.

Karl Adler

One of my main shareholders was a man named Karl Adler who worked as a private home decorator doing wallpapering, painting, and fixing up homes in the Detroit, Michigan area. I understood it was a very successful and lucrative business. Karl was very happy to invest in my Lucky Guy Company, as it would give him a chance to come into Canada.

Through Karl's home decorating business, he had great contacts in the Detroit area. He would mention to his friends that he had invested in my uranium company in Bancroft, Ontario, and he would line up investors to talk to me. Because of him and his contacts, it wasn't necessary for me to do much advertising for new investors into my Lucky Guy Exploration Company.

Karl had a friend named Jack Roy who owned a chain of hardware stores in the Gratiot area of Detroit. One day he said Jack wanted to become an investor in our company, and he wanted me to go to Detroit to meet him in his store. Since Jack was such a good friend of Karl's, I agreed to go to Detroit to meet him.

I met Jack in his hardware store, and we got along just fine. He said the most lucrative product he sold was paint that he created and mixed himself and that Karl was a good customer.

At one point Jack asked me if I could wait a few minutes while he did something. He went down to the basement and came back up with a bundle of bills and said, "Here's $10,000 I want to invest in your company."

When I looked at the money he gave me, the bills turned out to be the old-fashioned American hundred-dollar bills, and they smelled of mothballs. I was quite surprised and tried to figure out how on earth I could exchange these bills into Canadian money. Jack said he was unable to convert the money himself.

As I thought about the money, I called my bank manager and friend, Garvey Shearon, to see what he said about converting this money. He said, "Peter, just bring the money to me, and I'll take care of it."

When I took the money to Garvey, he too was surprised, because these bills were at least twenty or thirty years old, maybe even older. "However," he said, "it is good currency."

He put the money into my bank account $1,000 at a time so as not to draw too much attention to the old bills. Just by looking, anybody could tell that the money had been hidden away, and it was not very likely that there had been any taxes paid on it. Garvey never got over seeing that much old-fashioned American currency that smelled so badly!

Karl was a very interesting person. One time we were lost in the woods and came upon an old cabin. It was cold and raining, and I said, "Karl, we can't just break into the cabin. But if we could find a key we would be all set."

Karl said, "There's a key around the back, and I'll find it."

He proceeded to walk around to the back of the cabin and pulled a key from a hidden ledge on the second windowsill, and we were able to get inside the cabin.

We started a fire and were able to get warm and dry, and we ended up spending the night there.

Later, I asked Karl, "How in the world did you know where the key was hidden?"

Karl said, "It was part of my military training; we were taught to use our senses and intuition in these situations."

He didn't elaborate at that time and never talked about his life before we met.

From his accent, I knew Karl was German, but it wasn't until the end of our adventures together that Karl proceeded to tell me that he was an SS officer in WWII and that at the end of the war he was stationed in Norway. He related that the war ended so abruptly he had only his SS uniform with him. He said he had to steal some civilian clothes in order to avoid being detected as a member of the hated SS. He never related any stories of what he did during the war, and to tell the truth, I never wanted to know, so I never asked.

Karl did tell me that he survived between the end of the war in 1945 and his arrival at Ellis Island in 1952 by working in the black market. He did say though that he attempted to enter the United States three times before he was accepted in 1952 and then spent a year on Ellis Island.

Once in the United States, Karl had to think about how he was going to make a living. He couldn't take the chance of being on somebody's payroll, because there was always the chance someone would check out his past and discover his history. Therefore, he settled on being self-employed. He showed me the number tattooed under his arm that proved he was SS.

In the end, Karl bought the existing uranium property, and it turned out to be a good deal for both of us. He had invested in different ventures of mine. Eventually he and his wife bought a cottage in Bicroft near Bancroft. This was the name of a settlement built for the miners in the uranium fields, and when the mines closed, the cottages came up for sale. Karl bought one of them, along with some mining property, and he and his wife retired there.

I could never reconcile the Karl I knew with the Karl who had been an SS officer. I never did learn whether or not Karl Adler was his real name.

Uranium Mine Number Three

I started to work on a uranium deposit that I knew of up in Hershel Township in the Halliburton area.

With this property, I formed a new company and began to advertise, again encouraging American investors to meet me in my home in Windsor. With my den as a background, mineral samples, and knowledge of uranium deposits, it wasn't long before I had enough investors and finances to proceed with the work on this property.

We needed some kind of building, cook shack, and living quarters on this property. One of the investors from Lake Orion, Michigan, was connected to a trailer housing company. I told him I needed a fully loaded sixty-foot mobile home. He said he would sell one to me in exchange for shares in my company. I agreed and arranged for him to ship the trailer from Lake Orion, Michigan, to Wilberforce, Ontario. The trailer had all its American papers with it.

I made arrangements with the customs agent in Peterborough, Ontario, to allow the trailer into the country without my having to pay the taxes and duty. He agreed because I said it would be coming back out of the bush within six months and would be returned to the United States.

I hired a man named Vince Godfrey in Wilberforce who had a bulldozer. Using his bulldozer, Vince was able to make the road wide enough to pull the trailer right onto the property where we wanted it.

We had a generator running the trailer with all its electrical components. On the inside we used propane to run the stove and fridge. We now had a complete housing unit that was very comfortable for the workers and investors to use when they came to the property to inspect it and check how things were going.

The trailer was ten miles in the bush, and anyone who really looked at it would know that it was there to stay. After six months, I was notified by customs that the trailer was due to be returned to the United States. They wanted the trailer to clear customs in order to complete the paperwork on it.

I drove to Peterborough and met up with the customs agent. I explained to him that the trailer was so far in the bush that it was impossible to bring it out again. He said that he needed to see the trailer himself and that he wanted to make it a day's outing so he could get out of his office. He would be willing to come and inspect the trailer in the location it was in, and then he could make the determination of what to do with it.

When the agent first got into my car, he said he'd love to have a beer. I told him there was a cooler with ice-cold beer in the trunk of my car and that he was welcome to drink the beer on our trip to the property. The agent said that was terrific and that it would make the trip very pleasant. It was a three-hour trip from Peterborough to the property.

I could only drive my car so far into the bush, and the rest of the way I had to use a jeep that I had stored at a fishing lodge on Farcar Lake, which we had to pass on the way in. We got to the property about noon or shortly thereafter. The agent said he was glad he had come to see where the trailer was situated. He also said he understood that the trailer was so far into the bush that it couldn't be brought back out, because there was no access road, which meant the trailer would now be classified as being immobile.

Vince was at the trailer when we arrived and while the agent and I sat there watching TV, he made us lunch and the agent just kept on drinking the cold beer in the fridge. Finally, he said he'd seen enough. He restated that the trailer would now be classified as unmovable and that the value of it was zero—it now had no value.

When I took the agent back to Peterborough, he told me that it had been a terrific outing and that he understood that I was a very honest person and realized that everything I had told him was absolutely true. My company could have been fined and penalized for not returning the trailer to the United States, but the agent said it was trapped in the bush and all hell couldn't bring it out.

Garvey Shearon

Garvey was the best bank manager I ever knew. I first met Garvey when he was the assistant bank manager of the Bank of Montreal in Windsor, where I lived.

Garvey was always good to me and worked along with me in regard to anything having to do with banking. It seemed that he took a liking to me because of all my adventures and the different things that I had done and was doing in addition to working on the railroad. While involved in all my various adventures over the years, I was still employed by the railroad.

Garvey was the type that if you took a check to him to cash, he would cash it on your word if he knew you. Because he knew me, he was very easy going with all my banking business.

In Windsor, when we first met, I was having some difficult times, and sometimes I would need to cash a check in order to put the money into my own pocket to carry me over weekends and/or holidays. Anything I said I was going to do, I would always make sure that I did it and completed the agreement in regard to the checks, overdrafts, etc., that Garvey handled.

Garvey was transferred and promoted from assistant branch manager in Windsor to being bank manager of the Bank of Montreal at the East York branch in Toronto, Ontario.

If I needed anything, it wasn't necessary for me to talk to Garvey in person; I could just do it over the telephone. Things worked out very well for me by dealing with Garvey and his bank in Toronto. There were different money deals that represented large amounts of cash. There were times when my account was overdrawn and I was having hard times financially. Garvey would always cooperate to the best of his ability with any overdrafts, etc., that I required.

He and I got along very well together, and whenever I was in Toronto, I would visit with Garvey, and we would go to lunch or dinner together. Garvey was a WWII veteran. He had been a tail gunner and he told me he had completed twenty-five missions over Germany, after which he was eligible to be reassigned. Instead, he signed up for another tour of duty and did twenty-five more missions. Other than telling me this, he didn't discuss his war service with me. However, I did realize that he had a terrific war record and that any idiosyncrasies he had should be ignored.

Whenever we went to lunch, I would ask him where he would like to go, and he would pick a spot where there was lots of liquor. He insisted that I order whatever I wanted, and he always said he would just have a couple of drinks and then order his lunch. In reality, he drank his lunch and never ate anything, saying he wasn't hungry. He was the type that the drinks did not show or influence his condition at all. It seemed that his ability to consume alcohol was limitless.

One day, Garvey said he wanted to go to Bancroft, Ontario, with me. He had a week's vacation coming to him and he thought he would like to accompany me on one of my trips up north and spend his week there. I told him he was more than welcome to come with me anytime. I did go to Bancroft often because of the work and related things I needed to do on the property, and I was usually there for a week at a time.

I called Garvey one Friday and said, "I'm going to Bancroft on Sunday. How would you like to go with me?"

"Okay, I'd love to come. I'll be ready for you to pick me up on your way through Toronto."

I picked Garvey up that Sunday afternoon as prearranged and took him with me to Bancroft. I told him I always stayed at the Bancroft Hotel, and he said it would be fine with him. The hotel was laid out with all the rooms on the second floor, but there were no bathrooms in any of the rooms. You had to walk from your room down the length of the corridor to the one bathroom that was there for guests to use. Garvey and I checked in to separate rooms late Sunday evening.

Garvey said, "I want some liquor, Peter. Can I get it easily here in the hotel, or do I need to send out for it?"

"I have a couple of bottles of liquor sitting in my car right now if you want it."

"I'd appreciate it if you left a bottle with me. I need something to tide me over until I can get some more tomorrow."

"When I get up in the morning, I'll take you down for breakfast at about 7:30 a.m., and then after breakfast I'll be going out about 8:00 a.m. I'll give you a holler in the morning, and remember, all the time you are here, everything in the hotel is paid for, and that includes anything you would like to eat in the coffee shop. The property I'm working on is east of Bancroft, and you are welcome to come with me, but I'll be gone all day."

"Okay, I'll see you in the morning."

The next day, at about seven thirty, as I was headed for the coffee shop, I knocked on Garvey's door and told him, "I'll be leaving soon. Do you want to come with me?"

"I would prefer to stay right here and take it easy, so you go ahead and take care of your own business."

I didn't get back until about five in the evening. When I got back, Jack, the owner of the hotel, asked about the man I had with me.

I said, "Garvey Shearon is a good friend of mine. You don't have to worry about him, because he is my guest and I'll be taking care of everything."

Jack then proceeded to tell me, "He's got all my girls scared shitless. He just wraps a bed sheet around himself when he goes up and down the hall to the bathroom. In addition, he leans over the railing and hollers at Helen, my main waitress, to call him a taxi, and then he orders the taxi to go to the liquor store for him. The girls can't understand why he's never dressed and doesn't seem to want to come out of his room."

"Garvey is a very respectable person, and in fact, he's a bank manager from Toronto. Tell your girls they won't have any kind of trouble from him and just give him whatever he wants."

I was surprised Garvey was creating such a stink.

When I knocked on Garvey's door that evening, there was no answer, so I figured he was sleeping and just went about my business.

This scenario continued daily from the day we checked in until the following Saturday when we checked out.

All that time Garvey never got dressed, and he wore a bed sheet every time he came out of his room. He was the talk of the hotel employees because they had never seen anyone like him.

When we got back to Garvey's home in Toronto, he said, "Pete, you've got to come in with me. Thelma will want to meet you, and you can have a drink with us before you leave for Windsor."

I went in with him and met his wife. All the time I was talking to her, all I kept thinking about was Garvey in his bed sheet. How in the world was he going to tell her what he did on his vacation?

Thelma asked Garvey, "What kind of time did you have?"

"I had a terrific time. It was the best holiday I've had for years. I saw everything and did everything I wanted."

I was sputtering but didn't say anything to Thelma other than that I was glad Garvey had come with me and that he'd had such a good time.

That was the only holiday Garvey ever took with me, and he was the talk of Bancroft for weeks after. Everyone there called him "the ghost in the bed sheet". The girls never went into his room to change the bedding because he was always wearing them.

Even I was amazed he never got dressed the whole time he was in Bancroft. As I recall, he didn't eat much food either.

Garvey Shearon's Hideout – The Bancroft Hotel

The Priest

Some of the people you meet when you are running a company can vary from very strange to laughable. I recall one couple who had invested in my uranium company. They owned a banquet hall in the Van Dyke area of Detroit, Michigan. Their son came to Bancroft to check out the operation, and when he returned, the couple called me and said they would like to invest more money in the company. They wondered if I could come over and see them.

I said, "Sure, what's the best time and day for me to come?"

They replied, "You know we have the banquet hall and are always busy during the week. However, Sunday morning is our best time. We usually have coffee and breakfast, and it would work out great if you could come at that time."

I asked, "How much additional money are you looking to invest?"

They replied, "We're looking at $5,000."

"Of course" I said, "I'll be there at 9:00 a.m. this Sunday."

So over I went.

It was during the winter, and it was a snowy, damp, wet, and slushy day. When I arrived, they were sitting at the kitchen table in their banquet hall, and there was a priest and a neighbor couple sitting there with them.

The neighbors said, "We came to meet you and hear more about your uranium mine."

When I saw the priest sitting there in his frock, I thought the son of a bitch had come to divert my money into his church or something. That was the thought that was going through my mind, but I was all smiles.

They introduced the priest, and everyone called him "Father."

I said, "You look very good today, Father, bright and fresh. How come you are up and around at this time of the morning?"

"Oh," he said, "I have certain people that I'm obligated to bless every Sunday morning, and I always make sure that I'm available, ready, and willing to give them their blessing. I try to help them handle the trials and tribulations they may face during the week."

"Well," I said, "you're a wonderful person, and I'm sure the blessings you give your congregation really help and are appreciated."

The way I was talking to the priest gave the couple and their neighbors the impression I was a strong Catholic. However, we didn't discuss religion at all. The Father was sitting back from the table with his legs crossed, a hole visible in the bottom of one shoe, and he didn't have galoshes or boots with him.

I said, "Lord, your feet must be getting wet and cold."

He replied, "I don't worry about them, because I have more important things to do. Besides, I use paper to cover the holes."

I said, "The weather outside is so miserable, and I'll be leaving here shortly, so I'd be glad to drive you. We can't have you walking around with a hole in your shoe."

He said, "I'd appreciate that, and I'll gladly accept your offer."

The couple I came to see gave me a check for $5,000, and the neighbor couple said, "We have heard you are such a nice person. We've also heard about your property, and we would like to invest $2,500."

They suggested that they were doing it because we were all good Catholics and should all prosper together.

I told them, "We certainly will with the father's help."

After finishing his coffee, the priest said, "I have to take a minute to bless these people, and you are included if you want to be present."

I said, "That's very generous of you."

He blessed all of us at the table, and I noticed that the couple who owned the hall slipped him a hundred-dollar bill. The neighbor also gave him some money.

The priest was finally ready to leave and said, "I really would appreciate it if you could drive me since I have five more stops to make and my feet are getting quite cold and wet."

During our previous conversations, I had learned that he was from Windsor.

I said, "When we're done, I'm heading back to Windsor, and you can come with me."

So he gave me the directions to the five stops he had to make. Again, the people at these places had to be blessed. At all the stops, the people gave him about a hundred dollars or more in cash.

When we finished, he said, "I really don't live in Windsor. Right now I'm living at the priests' retreat in Wheatley, Ontario. It would be great if you could drive me there. First, though, I want to stop at a bar in Windsor."

He had to have a couple of drinks and buy some liquor to take back to the retreat with him.

We went into a bar, where he had his few drinks and got a bottle of whiskey, and I drove him to the retreat on Highway 3 on the edge of Lake Erie. It was only a couple of miles east of Wheatley. He and I were on very good terms and were very friendly during the trip home.

After dropping him off, I rehashed the day's events in my head and realized the priest's shoe with the hole in it was a prop. Every Sunday he would follow a predetermined route to visit people who wanted him to stop by and bless them. When having his cup of coffee, he took pains to let the people see the hole in his shoe. He was getting at least a hundred dollars or more a crack in order for him to have cash money during the week. He would end up with about $1,000 cash or more in his pocket at the end of the day.

To me, the priest was no more than an alcoholic con artist. I thought the retreat in Wheatley must be full of priests that the Church wanted out of the way until they could figure out what to do with them.

I was glad I had been so nice to the priest, because he was very important to the people who were buying shares in my company. They were all good Catholics, and it didn't hurt if I gave them the impression I was as well.

With this company, I had the opportunity to sell my shares to the board of directors. I didn't receive a large amount of money, but I did make a good profit and was glad to be free again to go on to another adventure.

Garnets

I knew there was a garnet deposit on Fishtail Lake, north of where I was, and I took Ted Boomer, my old friend and sometimes partner, with me to look at the property. Ted felt it was a terrific find. The garnets seemed to be a good quality, deep red in color, and some of them were suitable for gem material.

Ted, Cecil Dwyer, myself and one other workman began to use the capo drill to drill and blast the formation. The dike was full of garnets, and the formation was along the north bank of Fishtail Lake at the water's edge. When we were doing the drilling and blasting, we were partially under the water at the edge of the lake. After the blasting was over, we could see nothing but dead fish floating in the lake.

On the southern part of Fishtail Lake was a person by the name of Germaine. I'll never forget him. He tried to kill Cecil, Ted, and me.

Germaine was developing cottages and selling lots on the south portion of the lake. The one time we met at the boat dock he was very ugly and unfriendly to the four of us who were there. He was trying to say that we had no business drilling and blasting at the edge of Fishtail Lake and killing the fish. He vowed to stop us.

Later that summer, Cecil and I, about midnight, were driving our half-ton truck with a load of dynamite and gasoline. The dynamite caps were in the glove compartment. We were taking the supplies over the old bush road that led to our property. It was a clear, moonlit night, and Cecil and I were drinking beer and driving slowly over the bush road. We came to the top of a hill, and I told Cecil, "We have to stop here; I need to take a leak. It's such a nice night that we don't have to rush in and out again."

We both got out of the truck, and while we were standing there, I saw something shining across the road just ahead of us. I said, "My God, Cecil, do you see what that is?"

It turned out to be a quarter-inch steel cable strung across the road. If we had kept on driving and not seen the cable, it would have cut the top portion of the cab right off the truck and seriously injured Cecil and me. I said to Cecil, "The only fellow I can think of that would do something as dangerous as this is that damn Germaine."

The next day, I went to the Ontario provincial police in Bancroft to report what had happened. The officer said, "There have been lots of complaints about you and your crew dynamiting along the north shore of Fishtail Lake. We're waiting for an opinion from the fish and wildlife department to rule on your activities to see whether they are within the law."

Over the next few days, Cecil and I spent our time drilling under the lake as far as we could and setting off the dynamite. Finally, the game warden came down to the property to pay us a call. I showed him what we were doing as well as the garnets that were just glistening under the water. He understood why we were blasting under the lake.

"However," he said, "if you keep your drilling and blasting on dry land, you have full permission for that, because you hold the mineral rights on the property. But you cannot be drilling and blasting under the water and killing the fish."

We agreed to do as he said.

Later that summer, we sold the operation to an investment club from Detroit, Michigan. The day the representatives from the investment club came, the sun was shining, and when they were looking at all the garnets, the crystals were lying just underneath the water, just flashing and sparkling in the sunlight.

They agreed then and there to buy the property.

The Bancroft area is the most interesting mineral and gemstone area in the world. Universities bring their students to this area on field trips every year because of the numerous varieties of minerals and gemstone deposits. It is a Mecca for lapidary lovers.

Incidentally, this is the place Ted took a picture of me drying out my socks and boots. When we got the pictures developed, my mother was looking at them and swore there was a bear sitting behind me in the bush, just watching us. The bear made no sound, so Ted and I never realized it was there until mother pointed it out. Do you see a bear in the following picture?

Do you see a bear in this picture?

Graphite

Also, during my time in Bancroft, I worked on a graphite deposit in a different location and a mica deposit over in Eagle Township. Graphite was in great demand after the war. In addition, I worked on a beryllium deposit up in the Mammoth area. Beryllium is a gem material used in the space program for the nose cone of the satellites.

I had a museum-quality ore sample of 90 percent pure graphite weighing over seventy pounds that I had found on my property. Believe it or not, I hand-carried it over three miles out of the bush. It seemed like a great idea at first but quickly turned into a nightmare, but I was determined to get it out of there.

Confessions of an Eccentric Dreamer

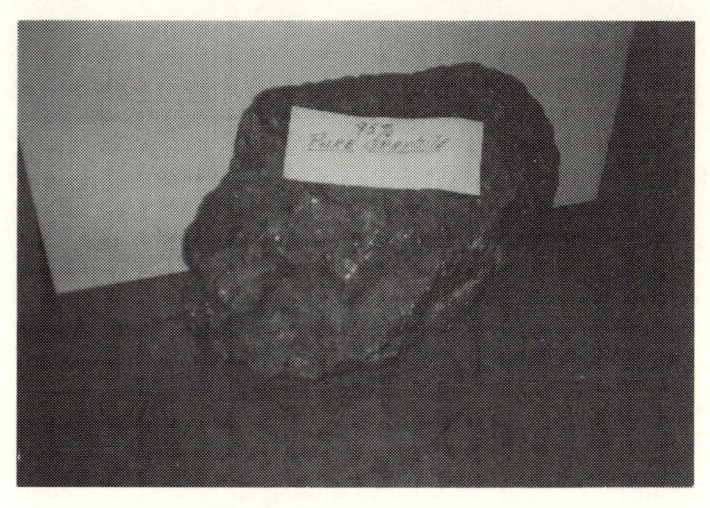

Museum quality Graphite

Rags to Riches

After the uranium and garnet claims were gone, I got gold fever.

Gold has many amazing qualities. It can be cast, carved, pressed, rolled, polished, or hammered the thinnest of any metal (as thin as 1/100,000 of an inch). It is nineteen times heavier than water, but it can be hammered so thin that it almost floats.

Gold has the most ancient history of all metals and was used for jewelry, ornaments, cups, and even burial tombs as early as 3500 BC. Egypt's young Pharaoh Tutankhamen was buried in a nest of three coffins. The innermost was solid gold, and the outer two were hammered gold over wooden frames and weighed 240 pounds.

Columbus and other explorers were seeking gold more than any other treasure. Early Spaniards in their lust for riches shipped loads of pyrite (fool's gold) back to Europe only to find truth to the saying *"All that glitters is not gold."*

The more I read and learned about gold, the higher and higher my gold fever got.

I had read through all the different gold reports and ended up staking property about 125 miles west of Cobalt and New Liskart, Ontario. I spent one summer working on this property that was next to an Indian reservation at Shining Tree. For me to even get to the property, I had to drive through Elk Lake and Cobalt over a washboard-rough road for about one hundred miles.

I had a couple of people working for me, but the gold property was very spotty. It had a pocket of very high-grade gold here and there, but every time I drilled and blasted it open, the gold seemed to disappear. I wasn't making any headway with it, but I kept thinking I was going to uncover some rich, visible vein of gold ore in order for me to sell the property. It was expensive, and all the costs were coming out of my own pocket.

How quickly things can change. The Indians I had working for me always wanted me to go and stake some of the rich silver mines that they knew about over toward Elk Lake and the Cobalt area. I kept telling them I was not interested in silver mining. I had gold fever, and that was what I was looking for. Just gold! Because of this, I ignored all their tips.

The price of silver at that time was approximately sixty to sixty-five cents per ounce. It was because of the low silver prices that there were many silver properties available and inactive. However, the price of silver then went up over a dollar an ounce, and all the silver mines started to be reworked. I had not staked any of these silver claims.

The thing that bothered me most about having the Indians working for me was that all you would hear every morning was they were waiting for the "bus." The bus made regular daily trips between Cobalt and the area of Shining Tree and Gowgamma, bringing liquor and beer supplies for everyone along its route. I got so sick of hearing about this damn bus that I made a trip to Cobalt and bought twenty-four bottles of cheap wine, twenty-four cases of beer, and twenty-four bottles of whiskey.

On my way back to Shining Tree, I was driving over this really rough gravel road, and even at twenty-five to thirty miles per hour, it felt like the car would shake apart. The driver's door lock came loose, and the door wouldn't stay closed.

I stopped at the Ford dealership in Elk Lake and asked the fellow in charge if he would come out and tighten the door-locking device on the Cadillac. I told him he would need a large Phillips screwdriver to fix it. He said he would be glad to help. He came out with the screwdriver and tried to get the screwdriver into the slots on the door, but his hands were shaking so bad he couldn't do it.

I asked him, "What the hell's the matter with you?"

"Well," he said, "I have a terrific hangover, and I'm waiting for the bus."

"God almighty, not the fucking bus again!"

"I drank all my liquor supply last night at a party, and there's nothing left to drink."

"Well," I said, "take a look in the trunk and on the backseat of this car."

I showed him the car was loaded with booze, and he nearly fainted when he saw it.

"My God, can I buy some off you?"

I told him, "Listen, fellow, there's nothing for sale, because I'm not in the bootlegging business."

His jaw dropped, and he nearly started to cry, but I said to him, "I'll tell you what I can do. You pick out whatever you want, and I'm going to give it to you for nothing, but don't take too much, because it has to go to the Indians at Shining Tree."

He picked out a case of beer and said, "I'd feel a lot better if I could pay for it."

I refused to take even a penny for it.

"I now have enough to get myself squared up. I really appreciate it and don't know how I can ever repay you for saving the day."

I didn't know this man, never knew his name, and never expected to see him again.

I continued on to Shining Tree, unloaded the liquor supplies in the chief's cabin, and told him to dole it out in small amounts. I kept working on the property until in the month of August I ran out of money and had to stop the project.

I came home to Windsor, got Marion and the two boys, and took them to Toronto, where we stayed at the King Edward Hotel. The Canadian National Exposition was on, and I wanted them to have a bit of a holiday. It would give me a chance to go to the mining recorder's office and investigate what kinds of properties were available that I might be interested in.

When I got back to the hotel from the fair I stopped in the hotel bar for a couple of drinks. I just sat there trying to think of what to do next.

I had just talked to my friend Garvey Shearon, the bank manager, and he told me that my bank accounts were overdrawn and that I had passed the limits in what he could let me have. He said he would appreciate it if I could help him out by getting my accounts back into a little better standing before they drew the attention of the bank auditors.

I finished my beer and was just going up to our room to get Marion and the boys when a stranger sitting at a table across the room came over.

He asked "Can I buy you a beer?"

I said, "No, thanks, I'm not in the mood to be drinking with my wife and children with me, so I really haven't time to enjoy your company."

"Well," he said, "I want to ask you a question. Were you ever in Elk Lake in a Cadillac that was filled with liquor and a loose driver's door lock?"

"Yes, but that was a couple of months ago."

"I know, and I'll never forget what you did for me. I'm the fellow with the hangover who was trying to tighten your door lock. You gave me a case of beer, and it really saved the day, and I have never forgotten it."

He continued, "It was such a surprise to me that I'd never before seen a car so loaded with liquor, wine, and beer before."

It turned out he was the owner of the Ford dealership where I had stopped, and he said he'd never in this lifetime forget what I did for him.

"Now," he said, "I think I have something for you. I happen to know that Marie Curie's uranium mine is open in Cardiff Township up by Halliburton. If you go right to the mining recorder's office, it can all be verified."

It was the kind of tip that you just had to check out. I immediately went to the Department of Mines, where Jack Andrews was the mining recorder. I checked in the library and found out that the mine was indeed open for staking. It represented approximately five lots on the tenth concession of Cardiff Township. I told Jack I had the property staked and needed to make out the papers that afternoon, which I did.

I finalized all the paperwork, took Marion and the boys back home to Windsor, packed all my equipment, and headed back to the Bancroft area.

Confessions of an Eccentric Dreamer

I called Cecil Dwyer, whom I had worked with before. Cecil was one of the most reliable workers that I knew. He went down to the mine with me, and in two days, we had staked all the property.

I was shocked and excited when I first saw the property because of what was on it. It had all the head frames and equipment, as well as three large buildings that were filled with the mining equipment needed to run the uranium mine.

I phoned A. J. McKenzie, a mining engineer with an exceptional reputation who was working on a nickel mine in Sudbury at the time. I told him I wanted him to fly to Bancroft right away to do an engineer's report on the property I had just secured.

An engineer's report is a bonded certificate of evaluation of the mining property involved. He did an exceptional report and described all the machinery and equipment, the great possibilities of the uranium mine being reopened, and its potential. The evaluation said the property and all the surface assets were worth about $750,000 and that I owned it completely.

I also phoned Garvey to tell him about the evaluation and to say that I had just gone from rags to riches, because I now owned Marie Curie's uranium mine. I then mailed him a copy of the engineer's report with a picture of the assets and equipment.

I was again set financially.

I never formed a company with this mine or sold any shares, so I owned it 100 percent. However, personal assets can change quite quickly.

Within a matter of months, Bancroft Uranium Mine sent a representative over to the Bancroft Hotel to talk to me about buying me out, which they did. I thought I would be able to get at least $250,000, but they wouldn't go that high. I needed the money to get back on firm footing financially, so I accepted a buyout of $195,000.

Some of you will wonder how the mine could be evaluated at $750,000 and yet be sold for $195,000. You must remember that an engineer's evaluation represents replacement value only and is not a buy/sell price. The sale allowed me to pay off all my debts, and it gave me enough money for my own personal use.

The reason that this uranium mine was so attractive to other mining companies was because of its history. Marie Curie in the 1930s had secured radium samples from this mine that she experimented with in Paris, France. She eventually died from radiation poisoning.

Marie Curie won the Nobel Prize in physics jointly with her husband in 1903. Later she won her second prize, in chemistry, and she was the first woman to win it on her own.

This was the hand of fate—for the cost of a case of beer. At the very least, it was the best investment I had ever made … and I didn't even know it at the time.

My first look at Madame Curie's Mining Property

Another Impaired Driving Episode

One morning about three o'clock, the roads were snow covered and slippery. I was on my way home from Bancroft and was driving too fast and sideswiped a car going in the opposite direction. There was a lot of damage to my Cadillac, and it was immobile. I thought that I looked a lot worse than I really was, because the cuts I had were bleeding.

My Cadillac at that time had what I called a knob in the middle of the steering wheel. This knob stuck out enough that when the accident happened, I hit my chest against it and fractured some ribs.

I had two or three cases of beer in the trunk and backseat of the car, and during the excitement, the first thing I did was to have one of the bystanders help me lift the beer out of my car and help me throw it in the ditch. I didn't want to get arrested for carrying open liquor in the car.

A squad car was soon on the scene, and I told the officer that I had to go to the hospital to be taken care of. I had cuts and my chest was hurting. He agreed, but he did mention that he could smell the liquor on me and would be charging me with impaired driving.

Before we left the scene, I looked for the beer I had thrown in the ditch, but it was all gone. One of the spectators must have seen us throwing it in the ditch and taken it.

On the way to the hospital, the squad car ran out of gas, and this threw the officer for a loop because of the delay in getting me to the hospital. When we finally arrived, nobody said anything about a blood test for drinking and driving. The policeman did, however, give me a ticket for impaired driving, even though I told him I was not the driver. I told him that because I was drinking, I had gotten someone to drive me home, but he had run away after the accident.

I wasn't seriously hurt, but I did have to spend a few days in hospital. I don't remember what injuries the people in the other car sustained, but I don't think they were hospitalized. I had to arrange to have my car towed back to Windsor. They charged me with careless driving, and I had to go to court in Peterborough, Ontario.

I found a lawyer by asking a taxi driver who was the best lawyer he knew of. He gave the name of a Mr. Chorley (or something close to it), so I went and hired him to represent me in court.

The main witness against me was a machinery salesman for a mining company, and he was someone who was following me down the highway just before the accident. The lawyer told me the court case was being heard after the lunch hour and was adjourned for two hours, so it gave me a chance to "soften up" the witness.

When I was going out for lunch, I told the salesman, who was very nervous and high-strung, to join me for lunch, and we went to the Peterborough Hotel. I knew the people in the tavern part of the hotel, and as we were going to have lunch, I told the witness to come and sit with me and that I would buy him his lunch and we'd quiet our nerves. I told the bartender we were going to eat later but to keep the drinks coming, because we needed a few to help us calm down.

By the time we got through drinking, we had to rush back to the courthouse, and we never did have time to have any food whatsoever. The court case started right away, and Chorley, my lawyer, called the main witness and put him on the stand.

He gave the witness the Bible and told him he "must tell the truth, the absolute truth, so help me God" during his testimony.

"Now," he said, "you realize that if you lie in the courtroom where you took an oath to tell the truth, you open yourself up to criminal charges and contempt of court."

While lecturing the witness, my lawyer had him so nervous he was shaking.

Chorley was the best lawyer I had ever seen in the courtroom. He had the machinery salesman so nervous he was sputtering.

The first question the lawyer asked the witness after the lecture was "Who was driving the car?"

The witness sputtered and stammered and said, "I don't know. McLean was there, but when I saw him he was in the passenger seat of the car."

The lawyer then asked, "Would you agree that he couldn't have been the driver of the car?"

The salesman replied, "No, he couldn't have been the driver, because he wasn't behind the wheel of the car."

The lawyer then asked the judge to dismiss the case against me because there was no proof of me being the driver of the car. The judge had to dismiss the case, but he was furious, because he knew that I had just gotten away with drunk driving and causing an accident.

It was, however, a good ending for me.

Never had I seen a lawyer lecture a witness to such an extent that the person was beyond nervous.

That was the one time a cab driver had ever given me such a great tip when he gave me the name of a lawyer to help me fight the impaired driving charge. It turned out that this lawyer was best known for handling impaired driving cases. He proved it when he handled my case.

WAWA ADVENTURES

Gold Fever

Once I got interested in gold mining, I turned all my attention to the Wawa, Ontario area. Again, I hit the books to study the government maps and reports before I went there.

As I found during my time in the Bancroft area, the Ontario government, through the Department of Natural Resources, has thousands of government geological mining reports covering all the areas of Ontario. One of the most important things for a prospector to do is to learn about all the reports available and how to secure them for examination. The government publishes a list of the available reports, and the majority of the reports are available through the mail.

During my research, I read the forty-forth annual geological report from the Ontario department of mines. This report described the Wawa area and in the report it noted that Deep Lake gold mine had an exceptional government report on it. Also, in the same township there were eight other past producing gold mines that were interesting and I found they were all open for staking. I made up my mind that my next prospecting trip would be to Wawa, Ontario.

At that time the Trans-Canada Highway was not completed at the Montreal River, and you couldn't drive directly to Wawa. They finally completed the highway in 1962 and I could then drive from my home in Windsor to Sault Ste. Marie and then straight to the Wawa area.

As soon as the highway was opened, I went to Wawa to start prospecting. I didn't have a company, and I didn't have any partners. Whatever I was doing was strictly on my own money. I began to look at all these past producing gold mines in the township, and I found Deep Lake, which I had previously read about.

When I checked, Deep Lake was still open for staking, and with the surrounding claims open, it made for a nice holding. I hired a fellow named Mickey Clement to stake it for me. I secured a group of claims including Deep Lake and had them registered. I was very happy about this.

I now had free reign to indulge my *gold fever*!

Clifford "Mickey" Clement

"The Diamond and Gold King" became Mickey's nickname after finding diamond samples, which he had confirmed by the government. This resulted in his name appearing in numerous newspapers.

Mickey is Wawa's greatest goodwill ambassador, booster, and promoter. He always feels that the next big gold or diamond discovery is just around the corner and will result in a bonanza for the Wawa area. He owns High Falls Lumber Mill, and between the mill, prospecting, and building log cabins, he is busy all the time.

I first met Mickey in early 1962 when I was in Wawa to stake a past producing gold mine that was open. When I got on the property I found out that Mickey had staked it the day before. I wanted to make a deal with him and tracked him down on an old bush road later that night.

Later, Mickey told me that when I first pulled up, he was thinking to himself, *Who the hell is this joker driving a Cadillac in the bush?*

Mickey and I talked and made a cash deal for the property on just a word-of-mouth agreement. Everything worked out fine. Ever since our initial meeting and for all the years that I've known him, Mickey's word is his bond. He is the most honest person you could ever deal with regarding any type of project he agrees on. Nothing in writing is ever needed.

In his younger days, Mickey was a trapper with miles of trap lines. He is the only one I've personally known who could go through the bush for a week with just a pocket full of teabags, a tin can, a jackknife, a wire snare, matches, and a small tarp. He could travel light and fast and live just by knowing what to eat and snare in the bush.

In his early years, Mickey worked for a large mining company in the wilds of Newfoundland before he moved to Wawa and settled there.

Over the years, he has become an expert log cabin builder. His cabins all had stone fireplaces that were perfect for heat and were smoke-free, a knack that not many contractors have achieved. He built all the log cabins for the Wawa Motor Hotel. Each cabin has a kitchen, fireplace, etc., and the tourists just love to stay in them. He also built all the cabins at the High Falls Tourist Camp, which he owned and his wife Dolly ran at one time.

Once, on a very hot mid-August day when the temperature was about 90 degrees and the mosquitoes were bad, Mickey and I went down to his High Falls Lumber Mill.

I said to Mickey, "It's too hot to work. Let's go out to the ski lodge and have some cold beers."

"We don't have to go anywhere, because I'm going to open up my own fridge."

He walked over to one side of the lumber yard and cleared away some sawdust. Underneath was ice from the prior winter with lots of bottled beer buried in it that was still ice cold. It was the best beer we had ever tasted.

The mill is located on High Falls Road and is miles away from any neighbors. How people knew what we were doing is anybody's guess. The first to show up was the game warden, who said he was just dying for a cold beer. Next came a diamond driller, Herb Funk, and his crew. The last to join us was a guitar player and his wife, who played a washboard. This is just an illustration of the mysterious ways that word travels in the *north country*.

We were all sitting around drinking the beer and talking about going to the ski lodge, but the game warden said he didn't have any money on him.

I told him, "You need to have some money, because there will be numerous people there, and it will be necessary for you to be able to buy a round."

"I'll have my wife bring me some money at the ski lodge."

I didn't think it would happen. However, when we got there his wife showed up and gave him $200, which surprised me. Never before have

I seen or heard of a wife who would bring their husband money to buy booze in a tavern—or, in this case, a ski lodge.

Herb Funk's wife, who joined us, hated the game warden because he had recently arrested their son and seized his truck and equipment for illegal fishing. Herb's wife called the Game Warden every name in the book and some that weren't in it. Even I was surprised at how vulgar she got, and I couldn't understand how he could endure such a verbal attack. I thought the game warden would get up and leave, but he just sat there and kept his cool.

He explained to her the rules and regulations in regard to hunting and fishing and how they were necessary to protect the wildlife. He started talking about the wildlife and their habits, and he kept us all spellbound with his stories.

I, for one, didn't know anything about bears, and the game warden proceeded to tell us their habits and movements and how smart they are. For instance, he told us about the bears going to the city dump at dusk every day to scrounge for something to eat. When they found food, they would take it, climb the trees, and eat it there. He had to go to the dump every evening, turn his spotlight on the dump, and blow his horn. All the bears would scramble out of the trees and make their escape using preplanned routes. The stories about the other wild animals were equally interesting.

After the first hour, he'd won Herb's wife over to his side, and she began to understand the reasons her son was arrested for his violations. We all gained a new respect for the Game Warden and his job. We all found him to be very interesting and honest.

I spent a lot of time with Mickey and his wife Dolly and he worked with me on most if not all my adventures in Wawa. We also got into a lot of trouble together. One time we got pinched because I brought my own whiskey bottle into a bar. For some reason, I was able to talk my way out of jail soon after, but Mickey had to spend the night there.

On another night, when we were sitting in a bar, I had a large roll of one-dollar bills on me. The more we drank and talked about how much money we were going to make with our gold properties, the more I thought we should share the wealth. I began throwing my one-dollar bills out the window and caused quite a commotion in the street with people trying to pick up the dollar bills. The bar owner told me to stop or get out. I stopped but still thought it had been a good idea.

Another time, Mickey actually saved my life. When I had claims on Mishibishu Lake, there were no roads to the property, so we had to fly back and forth by seaplane. On one of my trips back to Wawa with Mickey, after one too many drinks, I fell off the skid of the plane, and I ended up in the water underneath the plane and was unable to get up. Mickey had to dive into the water to save me.

When Mickey's son Clifford came to pick me up one morning, he found me lying unconscious on the floor of the motel. Because of my drinking the night before, Clifford at first thought I was just drunk but then thought it might be more serious so he picked me up and took me to the hospital. The doctor told me I nearly died and that I needed to have my heart checked out by a specialist. Needless to say, I ignored this advice.

Mickey wrote a book on building log cabins, and he built a classroom on his lumber mill property, to teach log cabin building. The government had a program whereby it paid unemployed men to go to school to learn a trade. Mickey taught the classes, but none of his students ever seemed to want to follow up and actually build log cabins. Eventually, Mickey added living quarters onto the classroom he'd originally built. He built a beautiful home where he and Dolly still live.

Nobody knows the Wawa area as well as Mickey's son, Gilbert.

To this day, Gilbert handles all my Wawa affairs. I just have to call and tell him the property location, and then he stakes the claim for me. We have a good working relationship, and I always transfer the funds to him the next day.

Gilbert is just as trustworthy as his father. It is a wonderful family trait. Both of Mickey's sons also build beautiful log cabins.

I could go on and on relating stories of Mickey and me and all the troubles we got ourselves into, but I'll stop here. Mickey and his wife, Dolly, are great friends. The people in the Wawa area referred to me and my friends as "the gold dust boys."

Down the road, I know Gilbert will help my son Guy and grandson Kai if and when they enter the world of prospecting.

Confessions of an Eccentric Dreamer

Top: Mickey and I relaxing in one of the log cabins he built
Bottom: Myself, Claus Becker, owner Wawa Motor Hotel
Dolly & Mickey Clement and Dr. Richard Dubinsky

Golden Goose Gold Mine

There was a lot of exploration work to be done at Deep Lake, and I knew that I wouldn't be able to afford it on my own. It was going to be very expensive to do the exploration and the cleanup that would be needed. The shaft had caved in and had to be re-timbered before we could enter it.

I came back to Windsor and formed the Golden Goose Gold Mine Company as a one-million-share company. I didn't want to have the company any larger, because I felt that by keeping the total number of shares at one million, it would be easier to sell shares in the company. So I began to sell shares, mostly to Americans, by advertising in US newspapers.

My lawyer, George Yates, formed the company and agreed to do the legal work for me in regard to further activities in the company. I had very good luck selling shares and raising money for the Golden Goose Gold Mine Company. A person from Windsor, Ontario, by the name of Jared Humphrey approached me and said he had heard about my gold company from one of his friends who belonged to a born-again Christian church in Detroit. Jared himself was a born-again Christian, and he gave me the impression that he was very religious and trustworthy.

Jared had the idea that he would be able to raise and sell shares in the company. His church members in Detroit, he said, were very interested, and that way, I would have some freedom from worrying about money and could have more time to work on the property.

Rather than give Jared a position on the board of directors, we made a deal that if he raised so much money—say, over $100,000—for the company, then I would appoint him to the board.

I've always felt strongly that it is important to have an odd number of people on any board of directors in any company. One of my original investors was a fellow by the name of Harold Stone. He was very easygoing and was a mellow sort of person who would do anything that another person wanted him to do. In other words, anyone could control him.

For my side, I thought putting Harold on the board was a good move and that the two of us working together could control the board. The company now had a board of directors that consisted of Jared Humphrey, Harold Stone, and me. I was president, Harold was secretary/treasurer, and Jared was vice president of the Golden Goose Gold Mine Company.

The work on the property was going well, and I had dewatered and re-timbered the shaft to make sure it was safe to enter the mine.

When I first went down the shaft, I saw the gold/quartz vein that was mentioned in the government report. It gave the impression that the old-timers who originally worked the mine were following this sixteen-inch-wide gold/quartz vein right down the shaft. They went to a depth of approximately 110 feet and then did some lateral work.

When we were first starting to work in the shaft after it was restored and made safe, one of the workmen I had working in the shaft was down about fifty feet. The shaft was on an incline, and you could scramble up and down the shaft, but we tied a rope around the worker for added safety. Soon he was hollering up to me that he didn't feel good and wanted to come up before he finished the cleanup work he was doing.

"What's the matter? Are you scared going down the shaft by yourself? Does it bother you?"

"I don't feel very good down there. I feel weak, and I'm having a hard time moving my arms."

"It must be your imagination getting the best of you, because being in the shaft shouldn't make you nervous. Okay, we'll bring you up."

When we got him up and settled down, I wanted to go down and take a look for myself to see what the problem was. Mickey Clement, who was working with me, tied the rope around me and lowered me into the shaft. I was down there about ten minutes when all of a sudden, I felt terrible.

I yelled, "Mickey, get me the hell out of here. I'm going to be sick to my stomach, and I can't move very well."

They quickly got me out of the shaft, and we sat around having coffee and smokes and tried to figure out what was causing the problem in the shaft. All of a sudden Mickey said, "I know what the problem is. The shaft is filled with dead air."

We had to have a blower system installed to blow fresh air down the shaft before anybody could work down there.

Dead air was something I had never heard of or thought was possible. Had we tried working without the ventilation system, we could have killed somebody. After we got the blowers working and filling the shaft with fresh air, everything was fine. That was quite a lesson.

Mine Shaft where we found "dead air"

Susan

I spent a lot of time in Wawa alone because Marion couldn't come with me. I have to be honest and admit it was quite lonely. There was a lot of drinking during these trips. I even resorted to placing ads in the newspaper for a female companion to accompany me on my trips. The ads said I was a non-drinker and non-smoker. Both were untrue.

At one point, I did find a friend who lived in Wawa by the name of Susan who owned a beauty salon. She was a very popular hairdresser, and the women in the area seemed to like her style. Susan was with me during many of my adventures in and around Wawa.

Her shop was located right next to the police station. In fact, it was actually on the same floor.

The police were always checking up on Susan. If we were sitting in the car, the police would pull up next to us and say something like "You have a problem with a headlight."

I always told them "I'll have it checked out and make sure it's working", and they would leave. Susan said they stopped just to see if she had been drinking, but I thought they were just checking me out to be sure we weren't sitting there fighting or arguing.

Susan did have a boyfriend, however, and one night while Susan and I were at the bar of the Lakeview Hotel, the bartender told me, "Peter, you'd better be careful because Susan's boyfriend has a gun and he seems to be all worked up and is searching everywhere for you".

We left there and went to the Wawa motel for a couple of beers. While sitting there, we thought the boyfriend would show up, so we left and went to another motel down the road on the outskirts of Wawa. I said, "I still don't feel very comfortable here, and I think we should leave."

We left there and went to another motel. We got undressed and into bed, but I still didn't feel comfortable, so we left there and drove to the Trail's End Lodge at Montreal River Harbour. It was off-season, so the room had been closed up, and it was very cold. We again got into bed, but it was still too cold, so we got dressed again and drove all the way to Sault Ste. Marie and checked into the Diplomat Hotel. It seemed like we had tried five different motels because we kept thinking something bad was going to happen.

We stayed in the Sault for a few days so Susan could do some shopping before we returned to Wawa.

The boyfriend was a pilot for the Wawa Air Service, which flew tourists to and from the area's hunting and fishing camps. When we got back to Wawa, we heard through the grapevine that he had been transferred to Chapleau, Ontario, so we felt from then on we would have some peace and quiet.

About four in the morning one day, I was leaving Wawa, and Susan was with me. There was a car ahead of us going slow, and I told Susan that as soon as the road straightened out, we'd pass it and be on our way to Sault Ste. Marie.

When we came to the next rise in the highway we saw that the car ahead of us had hit a moose on a bridge, and we stopped just in time to avoid smashing into them. I thought to myself how lucky we were that I hadn't passed that car, or it would have been us who hit the moose.

There was a family of four in the car, and they were from Ohio and just starting out on their vacation. The moose had gone through their windshield, and the smell from the innards of the moose was terrible. The car was smashed up so badly that it was unmovable. I told the wife who was all cut up from the shattered windshield that I would take her to the hospital in Wawa.

I wrapped the woman up in a blanket and put her in the passenger seat. The husband, the two kids, and Susan were in the back of the car. As I started to drive to the hospital, the woman started to scream and wouldn't stop.

I told the husband, "My head is ringing from the screaming and I can't stand listening to her".

No matter what the husband tried, she wouldn't stop.

I remembered that I had a gallon of vodka and orange juice in the trunk. I pulled over to the side of the road, got out and filled a glass with it and said to the husband, "Give your wife some of this to try and settle her down".

She did take a good size drink and finally started to quiet down.

On the way to the hospital, I stopped at the police station and told them, "A car has hit a moose and it's still sitting in the middle of the road blocking the bridge. I'm taking the passengers to the hospital in town and you need to clear the road before another accident happens".

Susan and I stayed at the hospital until we found out how bad the woman's injuries were. It turned out they were mostly shallow cuts to her face. The husband thanked us for driving them to the hospital, and we left and went on our way to the Sault.

There was no way to call for help because of where the accident happened, so we were glad we were there to help the family.

Susan was around for several years. Never was the subject of my getting a divorce brought up. I had made it quite clear from the beginning of our affair that I would not leave my family.

About five years after I first met Susan, she got married to a man who worked in the mines.

Marion

During one of my trips to Wawa when I was drinking heavily, I was gone about three months without calling home and Marion didn't know if I was dead or alive. She had tried to get me on the telephone but nobody seemed to know where I was. Marion finally drove all the way up to Mickey's (an eleven hour drive) and asked him to track me down. There I was drinking in a bar with Susan when Marion finally found me. I can't remember whether or not she believed my story.

Marion was one very smart woman, and I cannot believe she didn't know or suspect what I was up to when I was away from home for extended periods of time. Marion, however, always wanted to keep the family together and the family came first.

She wanted to know how it would look if she called the police to report a missing husband; one missing for over three months before it was reported. I agreed this could have raised questions, and they would probably have wanted to know why it wasn't reported earlier.

Needless to say, it was a very awkward situation, but Marion and I worked it out.

Another Trailer

Just like I had done while working in Bancroft, I bought a fully furnished and equipped trailer from one of our shareholders. The trailer contained everything we needed, including a television.

Mickey and I were going to take it and put in on the property in lieu of a cook's shack. Then, when we were there, we could live in the trailer and eat and sleep and be comfortable. All we needed was a generator to run the electrical system of the trailer and a couple of propane tanks for operating the stove and refrigerator.

Mickey thought the trailer was a great idea, but he didn't know how in hell we could get it onto the property. There was no passable road to the mine, and we had been using an old four-wheel-drive jeep at the time. Since I had run into the identical problem earlier in Bancroft, I knew how and what had to be done.

We needed a bulldozer to be able to move the brush, trees, and rocks to make a road so we could haul the trailer right onto the property.

While we were making our plans to put the trailer where it was needed, I discovered there was an artist staying in the High Falls Tourist Camp that Mickey owned at that time. This got me thinking about having a painting done of the mine and property as I envisioned it would look when the mine was up and running. I talked to the artist about what I wanted and he agreed to come with me to the mine to do some preliminary sketching. He could then come back out to the Tourist Camp to finish the drawings and complete the painting.

The morning we were leaving High Falls Camp to take the trailer to the mine, I picked up the artist and found his wife was hollering at him that he was not to be doing any drinking with us on the trip to the mine. She seemed like a very hard-bitten and strong-willed woman who could tell the artist what he could and could not do while he was with us. She reluctantly gave him permission to make the trip.

I asked the artist if his wife was bossy or if it was just my imagination. He said she certainly was and was always telling him what to do and how to do it. I told him he was in a hell of a fix with that kind of woman as his wife.

Mickey had gone ahead of us with the trailer and the bulldozer. I told Mick we would catch up with him easily and follow behind them.

I decided to use my old Volkswagen van, and between the front seats I had a cardboard box full of the groceries I had promised the workers I would pick up. In the box was a large bottle of "Home Ketchup," the cheapest I could buy.

We caught up with Mickey at the edge of Deep Lake. He was stuck with the bulldozer and trailer at the top of a steep hill when I came driving up behind him. The Volkswagen was on a backward slant, and I had forgotten it didn't have any brakes.

We started to roll back down the hill, and I had a choice of going into Deep Lake or swerving the van up onto the embankment on the opposite side of the road. I chose to swerve up the embankment, and we upset the van. When this happened, the big bottle of ketchup broke. The artist ended up sitting below me in the overturned van and the ketchup spilled all over him. The poor guy was covered in ketchup from head to toe. He was a real mess but unhurt. My friend, Susan, was also with us but she was not injured either.

Mickey finally cleared the top of the hill with the trailer and bulldozer and got a couple of the workers to help us lift the Volkswagen upright on its wheels. Everything was okay except for one broken side mirror.

The van was still running so we kept going, and finally we ended up at the gold mine with the trailer, bulldozer, Volkswagen, and an artist covered with ketchup. The artist looked like he was drenched in blood. We even managed to bang up the side of the bulldozer.

I told the artist to come and sit down and have a cold bottle of beer to sip on while he got his nerves back into shape so he could do the sketching I wanted done.

I said, "Never mind the ketchup on your clothes. It will have to wait until you get back to High Falls Camp, where your wife can get you a clean set of clothes."

We finally got the trailer situated and set up on blocks, and then we got the generator and propane hooked up. We then had all the comforts of home, even a television that worked in the bush.

I told Mickey, "That's enough for today. You take the bulldozer back, and I'll follow you in the Volkswagen, and we'll meet at the Lakeview Hotel for a few drinks."

We drove back to town and went into the Lakeview Hotel Tavern. Mickey, Susan, my girlfriend, the artist, a couple of others, and I were sitting at a table approximately fifty feet from the side entrance. Somehow the artist's wife found out where we were, and after three or four drinks, the wife stuck her head in the side door. She could see her husband and she started to holler.

"Oh, Jimmy, Jimmy, what's happened to you? Are you hurt bad? I'm so glad to see you in one piece. Where did all the blood come from?"

It shocked her so much to see her husband in the shape he was in that she stood and did all this hollering from the doorway. Jimmy said, "Never mind standing there, come and get over here and sit down and be quiet. We're busy drinking, and I don't want to hear any more squawking and nattering from you. We'll talk about it later."

I told Mick, "Boy, we've done the artist a tremendous favor, because he'll end up being the complete boss of his own family."

The painting the artist did turned out to be exceptional, and it was exactly what I had imagined the Golden Goose Gold Mine would look like with the building and head frame on it. I was extremely happy with the picture, but I'm sad to say that today I can't remember what happened to it.

It seemed that after setting up the trailer and fixing a road so that it was passable to the mine, we were involved with nothing but thefts. I even hired one of the workers to live in the trailer and stay on the property 24/7 as a guard. In the end, though, there was nothing left of the trailer—not even the frame.

Trailer brought onto property and how it was destroyed

Golden Goose Shareholders

Since I put Jared Humphrey on the board of directors, he was busy signing up aditional shareholders. Because of his efforts, the company was well financed, and I was riding along with full expenses and good wages. I had it in the back of my mind that I'd better take out a stock option in the company so that I would be able to control the company in the future.

Harold Stone and Jared Humphrey were getting very friendly with each other, and at the first meeting of the Board of Directors, I brought up the stock option I wanted. I didn't see any reason why I shouldn't have it. Jared Humphrey and Harold Stone voted me down. As mentioned earlier, the board of directors was made up with just the three directors, and the majority vote controlled the activities of the entire company.

After the meeting, Jared asked me if he could have a Dr. Jull, the head of the geology department at the University of Windsor, assign one of his honor students to come to the property and make out a geological report. I said I would be glad to do it, because an additional report would help establish the gold value of the mine and what we were doing.

The next time I was in Wawa, Dr. Jull arranged for a student by the name of Binny to come and do a geological report on the mine, and I booked him into Mickey's High Falls Tourist Camp.

I asked, "Binny, when do you want to go out to the property?"

He replied, "That's what I came for, so we should do it tomorrow."

I told him, "Okay, we'll leave the camp at 8:00 a.m."

At eight, I stopped by the camp to pick him up, and Binny told me to come in while he packed his lunch. His lunch consisted of a sandwich made up of half a pound of butter between two pieces of bread. I didn't pass any comment at the time, but on the way to the Golden Goose Gold Mine, we were talking, and he said he was from Pakistan and that this was his favorite sandwich to make for lunch: lots of butter between two slices of bread. I told him we had coffee and everything to drink at the mine.

When we got to the mine, I showed Binny where everything was located. I showed him the general outline of the property, the gold/quartz vein, the shafts, and whatever I thought he might want to look at. I told him that since he was doing this report for somebody else, I wouldn't interfere with him or be with him while he was doing his inspection.

So during the day, I kept myself busy doing other things, and I made it a point not to talk to Binny in any way while he was working. I wanted the report to be authentic, and I didn't want anything said in the future that I influenced the student geologist in any way. In fact, I didn't even offer him a beer.

We were on the property until four thirty in the afternoon, when Binny said that he had all the facts he needed and wanted to get back to the High Falls Camp so he could write his report up that night. I dropped him off and told him I'd come by the next morning to see how he was getting along.

When I got to the camp early the following day, Binny said the report was finished. I asked him if I could read it, and he told me I was welcome to see the report and look it over, which I did.

The gist of the report was that he had seen over a million dollars in visible gold on the property. I'd never seen such a report, and there was no other mining engineer or geologist who would make out a report with those statements. I was happy as a lark, though, and I didn't ask for a copy, because as I said before, I didn't want to be seen as interfering in any way. Binny said he was finished here in Wawa and was heading back to Windsor to give his report to Dr. Jull.

I felt the company was running smoothly with lots of money and thought this report should put it over the top with investors.

The Golden Goose Mine Company was a private company whereby you could only have fifty shareholders excluding the board of directors, and we were getting close to that number. If we went over that number of shareholders, we would have to change our status to a public company and proceed from there. This change would be a very expensive proposition.

When I got back to Windsor the following week, I asked Jared what he thought of the geologist report. He said it was tremendous and that he needed to have a meeting of the board of directors. I said that was fine by me and to let me know when it was scheduled.

We were meeting in George Yates's office and Jared said, "I want to have the first annual meeting of the company."

I replied, "It isn't the time for it since we have another few months before the due date for an annual meeting."

Jared said, "I want us to take a vote right here and now."

Harold and Jared voted in favor of the annual meeting.

I asked, "Where do you want to hold the meeting?"

They said, "We want it right now, right here, in George's office."

Then I told them, "There's not enough room here in George's office for all the shareholders."

Jared said, "All the shareholders are on my side and would be in support of anything I did, and they would not feel it was necessary to attend the meeting."

So we had our first annual meeting.

Jared said, "I have proxy votes in my briefcase. Along with these and Harold Stone's stock, I now have absolute control of the Golden Goose Gold Mine."

He turned to me and added, "I'm going to ask you to resign as president and from the board of directors immediately."

I replied, "What am I going to do with all my stock?"

"We're prepared to buy them, and you can consider yourself out of the company completely."

"Okay, if that's what you've got in mind. You've got the votes to do it, and I can't stop you, so it's okay with me."

George said, "What Jared Humphrey is doing is perfectly legal, and you have no choice, Peter, but to comply."

"Well," I said, "I've got to have cash money for the total amount of shares that I have in the company."

Jared said, "I've got the money available, and I'm making out your check right here and now."

I was dumbfounded, but I really didn't give a damn, because I knew the newest report was way out of line and that the company would soon incur the cost of having to be changed to a public company. Also, there are a lot of expenses involved in diamond drilling on the property before it would prove to be profitable.

As I walked out of the meeting, I was thinking how disappointed I was in Harold Stone. I'd thought Harold was my supporter in the beginning and that he would be loyal enough to stay with me. I thought he would stay on my side of the fence, as we only had three people on the board of directors in the company. With Harold's vote on the board, we could have controlled the company, and I could have done anything

I wanted to do until we had the first annual meeting. And, even at that time, I could have postponed the meeting for at least another six months. But that's not how it turned out.

This is an example of the different turns and twists and things that can happen to you in regard to having a report such as the geology student Binny made out. It brought out the greed in Jared Humphrey, and it showed his true colors. He said he was a born-again Christian, but as far as religious activities were concerned, he was a first-class scoundrel.

I came out of the company, however, very well compensated. In fact, I came out smelling like a rose.

I made sure to tell everyone in Wawa that I was out of the company and had nothing more to do with it. I shut down everything I was involved with in Wawa that had anything to do with the mine.

The Golden Goose Gold Mine was my first project in Wawa. I considered it my pet project as well, but it didn't slow me down. I now had lots of money in my pocket to run the house and other mining expenses I might incur.

Again I had my freedom to seek other adventures.

I must say the reason I called it the Golden Goose Gold Mine was because of the large goose that stands at the entrance to Wawa as a symbol of the town. I felt "Golden Goose" would be a good name for the company, and it was. It also seemed to have a lot of appeal to it with the shareholders.

A few months later, one of Jared's shareholders told me that Jared was barred from any contact with his church's congregation. The reverend and the people running the church realized Jared Humphrey was taking all this money from the congregation for himself when the money was supposed to be going toward supporting the church.

I also ran into and talked to Harold Stone one day on the street in downtown Windsor. I asked him how things were going, and he said everything had gone flat. There was no activity in the company at all.

Confessions of an Eccentric Dreamer

"Goose" at Wawa Tourist Information Center

Al Turcott

Al lived in Wawa and was a great visionary who wanted to improve the town and the life of its citizens.

Al worked very hard to get the Trans-Canada Highway finished between Montreal River and the town of Wawa. Previously, you could not drive a car directly from Sault Ste. Marie to Wawa.

I was one of the first to drive the new highway.

Al owned the Salt Mines Motel, where some of my workers and I would stay when in town. There was parking on the roof of the second floor of the motel, and I used that for my van, snowmobile, car, and other equipment so they would be safe from theft and vandalism. It did not have a fence or railing around the roof, but it did have a twelve-inch piece of timber around the edge to keep cars from driving off it.

One night, while a worker of mine was running my snowmobile too fast up on the roof, the throttle stuck open, and the snowmobile flew off the roof, across the parking lot, and through the basement window of the Lakeview Hotel next door. Fortunately, the worker was not injured.

I told Al what had happened and that I didn't want to involve the police, who had their offices right next door to his motel. I asked him to find out the cost of the damages so I could settle up everything. This I did, and Al always told me that he was always there to help.

Al previously owned the Wawa newspaper and was one of the mainstays of the town. He was well known and well liked.

Al was the sort of individual who was always thinking up new and exciting ideas. He owned about 240 acres down back of the mission on the Michipicoten River. He had secured the property by staking it as a mining claim and declared that he was looking for diamonds. This was before anybody in the Wawa area or anybody in Canada thought there might be diamonds nearby. He was able to have a special act passed through Parliament that he could patent the land and own it outright.

Al sold the Salt Mines Motel around 1969 and decided to build a complete Indian fort on the land he owned and use it as a tourist attraction. The main building was fixed up as a museum and shrine; a rock shop, a restaurant, a lookout tower, and a trailer park were also included. The fort was built in support of the Indians, and it was called "Fort Friendship" for their benefit.

There was always a lot of drinking at the fort and in the vicinity. It seemed that everybody who came by had a bottle of liquor with him or her. Al always pondered what he could do with all the empty bottles.

One day Al decided to build a church made up of empty liquor and wine bottles. When anyone came by, he would encourage them to donate bottles to the church, and he also encouraged people to put donations in the bottles.

During construction of the church, when someone donated money, Al would write their name in a ledger. Then they would take the cork out of a bottle, insert money and a slip of paper showing their name, and then replace the cork. He began in this way to gather the bottles, and before he was finished building the church, he had thousands of bottles built into the walls. As you can image, Mickey and I donated many of the bottles.

All the bottles had the cork/neck facing toward the inside. Al called it the Church of Departed Spirits. It was the only church most people had ever heard of that was completely built out of empty liquor and wine bottles. When the organ was playing, the sound echoed out of the

Confessions of an Eccentric Dreamer

empty bottles; it was very eerie and stimulating. It was also very strange and something I had never heard before or since.

Al also invented pointed bumpers for the front of cars and was trying to prove to automobile companies and anyone else who would listen that with this type of bumper, it would be impossible to have a head-on collision. He had two or three cars made up with plywood bumpers, and when we would go to the fort, he would want us to drive the cars around and crash into each other so that he could illustrate the principle of his pointed bumper.

I thought it was a tremendous idea. However, the automobile companies were not interested, because they felt it spoiled the look and design of the car.

Al needed publicity to attract tourists. He was invited to be on a television show in Thunder Bay to talk about his cars with pointed bumpers, but his main objective was to publicize his church. Through the advertising, he was able to finish the altar and add the pews.

He did have two or three weddings in the church, and he would have liked to have actual church services there, but the ministers in the vicinity each thought that it would be too disrespectful to be used as a house of worship.

Fort Friendship was never the success it should have been. It was located approximately seven miles from Wawa on the Trans-Canada Highway and was not advertised by the town of Wawa and tourist bureau as an attraction, as it should have been. The business people felt if it had succeeded it would have drawn all the tourists away from the town.

He tried to have signs put up along the highway to advertise the access road but the government would not give him permission to advertise the location of the fort. This was after Al had spent thousands and thousands of dollars building it. The fort was finished in 1970, and it was a wonderful tourist attraction to the many who visited it.

Al lost his health around 1972, and he died of cancer in 1974. He was sixty-eight years old. He wanted to be buried on the banks of the Michipicoten River near the fort and the church he had built. His headstone is located where he wanted it.

After Al died, Mrs. Turcott was unable to keep the fort and the church together. Over a two-year period thieves and vandals took or destroyed everything Al had worked so hard to build.

Mrs. Turcott had numerous ceramic wild geese made up and attached to a base as souvenirs, and I bought a box of the geese from her. In fact, I still have some in my possession. The company I had been promoting, Deep Lake Gold Mine, was renamed the "Golden Goose Mining Company." I had the geese painted a gold color and gave one to each of the shareholders.

I felt very bad about the destruction and vandalism at the fort, which was mostly the result of Al not being able to use highway signs to attract tourists. People wrecked the church, broke all the bottles, and stole all the donations that had been left. They also destroyed the buildings. It was completely wrecked.

I have always felt that if Al had lived in a large city such as New York, Toronto, or Vancouver, his deeds would have been rewarded and recognized. I also felt the citizens of Wawa should have done more for him.

Incidentally, it was Al who had the big white Canada goose built as a tourist attraction on the highway at the entrance to Wawa.

In my opinion, he should have been buried in Wawa with an appropriate ceremony, not isolated and lonely in a grave seven miles out of town.

Right: Author entering the "Church of Departed Spirits"
Left: Interior of "Church"

George Kenty

Some friends and I, including Mickey, were all spending an evening having a few drinks at the Wawa Motor Motel. With us was George Kenty, a mining engineer who did work for me. I was driving my car, and George was using my Volkswagen van. When we were leaving the Wawa Motel to go back to the Salt Mine Motel, where we were staying, the headlights of the van came on, and I assumed George was in the van and was okay.

As usual, I parked on the roof at the Salt Mine Motel. When I checked the next morning, the van wasn't where we kept it parked.

I went down to George's room and found him there and asked, "Where the hell is the van?"

"I left it out at the Wawa motel and got a ride back here last night. I thought I was too drunk to drive."

I said, "Well then, we'll have to go back out there to get the van and bring it back here."

I got my car and George and I drove back to the Wawa motel. When we got there we decided to stay for a few drinks in the bar.

When ready to leave, I got back in my car, George got in the van, and we headed back to the Salt Mine. I got there first, and after a couple of hours, there was still no sign of George or the van.

I got a call from the police and was told, "Your van has been in an accident and the driver is in jail."

When I asked if anyone was hurt, they said no and I immediately went over to the police station to see what had happened.

On the main street in Wawa, there was a traffic island in the center where cars would turn a small corner. George had run up onto the island and upset the van and the cops pinched him for drunk driving. The whole scenario was hard to fathom because it was such a bright, sunny day.

When I finally got to see George, the first thing he said to me was, "The lights weren't working on the van and that is why I hit the island".

It turned out that this was the same statement he'd given to the police. I asked him. "What difference would it make if the lights were working or not on such a bright afternoon".

He had no answer.

I bailed him out of jail. When his case came up, he lost his driver's license for three months.

There wasn't much damage to the van, and the policeman said that George's excuse for the accident was a new one for them.

I got the impression that if George had said nothing, the police would not even have bothered giving him a ticket.

Harold Newton Sr. and Harold Newton Jr.

I have previously told about an adventure on the *Marion B* that I had with Harold Newton Sr.

On another occasion, I needed to take my half-ton truck to Wawa along with a generator and other items needed for mining. Since the truck would be left up there, I needed someone to drive it up for me.

Harold (Newt) Newton was a machinist on the railroad who worked for me in the engine house. I asked Newt to drive the truck up north for me. He and his son, who was a real nice kid about twenty-three or twenty-four years old, and if I remember correctly, he was a football player with the Hamilton Ti-Cats, agreed to drive the truck for me. We had an uneventful trip up to Wawa.

On our way back to Windsor, we stopped for the night in a hotel in Sault Ste. Marie, Ontario, where Junior started jumping on his bed and hitting his head on the ceiling of the hotel room. About three in the morning, the hotel manager woke me up and told me that Harold Jr. was making too much noise in his room and causing a real problem with the other guests who couldn't understand what was happening. He asked us to please leave the hotel.

I said I would take care of it. I still don't understand to this day what had gotten into Junior or why he was making so much noise, but because of it, we got evicted from the hotel.

On our trip back toward Detroit, the three of us were moaning and complaining about our hangovers. We decided to stop in a bar and have just enough drinks to tide us over until we reached Detroit. About seven in the morning we came to Cleo, Michigan, which is about twenty miles north of Flint, Michigan. Junior was driving, and he pulled into the town because we thought there would surely be a bar open on the main street.

There wasn't a parked car anywhere to be seen, and there was nobody on the street, but there were numerous bars. I told Junior to let me out of the car and said I'd walk along the street and try the doors to see what was open. I tried about four bars, and they were all locked up tight. At the fifth bar, I found the door was unlocked.

I had Junior park the car, and we went inside. We didn't see anyone in the place. We hollered for the bartender, but there was no answer. There was not a soul there but us. I figured someone had just left for a few minutes and would be right back.

While we waited, we helped ourselves to beer and a couple of shots of whiskey. When still nobody came, I told Junior to check out the toilets and any other rooms to see if the bar had been robbed and the owner injured. He didn't find anything. We stayed for a couple of hours, helping ourselves to whatever kind of drinks we wanted, since everything was wide open.

At one point I got on the telephone and told the operator, "We're in a bar and there's nobody around, and we can't find anybody to pay for the drinks we're having."

"Do you realize it is Sunday morning and bars don't open until noon?"

"I didn't realize that, but whoever owns the bar forgot to lock it up when they left last night."

After hanging up the phone, I told Newt and Junior, "We'd better knock this off and get the hell out of here before the police come and accuse us of breaking and entering".

We left forty dollars on the bar with a note saying we had been there and had a few drinks. I made the note short and sweet and didn't sign my name. We got back in the car and headed for Detroit.

We laughed and laughed at how lucky we were to find a bar with an unlocked door so that we could have a few drinks to help keep us going.

What were the odds!

An Irate Husband

Bars in Sault Ste. Marie, Ontario, close two hours earlier than they did in Sault Ste. Marie, Michigan. When the bars in Canada closed, a large number of Canadians were in the habit of going across the border to continue drinking in Michigan. The Northview Tavern and Hotel was a popular spot to finish off the evening.

One night a group of us headed over to the Northview, and I kept dancing all night with the same girl. During the evening, she told me that she had had a big blow up with her husband and wasn't going to go home. She asked, "Where are you going to stay?"

I told her, "I'm staying right here in the Northview because I don't want to drive my car back to Canada in my condition. If you are stuck, you are welcome to stay with me in my room."

So she did.

Her first name was Darlene, and I will never forget her. The evening ended, and we spent the night in my room.

I woke up at daylight the next morning and said to Darlene, "There's nothing to drink, so I'm going to go out and get something. There's a bar just down the street that opens at 7:00 a.m. I'll go there and bring something back."

So I got dressed and went down the street and brought back a supply of liquor to the room. I opened Darlene and myself a bottle of beer, and we were just going to have a peaceful couple of drinks each to help us get squared around for the day before we left the room.

About ten minutes later, there was a knock on the door, and when I opened it, there stood a detective in a suit and tie. With him was a member of the US Border Patrol in full uniform.

I said, "Holy Jesus, what do you guys want?"

They asked to come in and when I let them in the room, Darlene was still in bed. The detective told Darlene, "You should be ashamed of yourself. Your husband has just spent the night looking all over town for you. He has a shotgun with him and is threatening to kill someone. Someone could have gotten hurt."

"Well," I said, "Darlene asked to stay in my room because she didn't have any other place to go."

They knew that she and her husband owned a business in Sault, Michigan. They also said to me, "Her husband is very hot-tempered, and he's been out looking for someone named Peter who drives a white Cadillac."

They wanted me to hightail it back to Canada, and they wanted Darlene to get out of bed and get dressed so they could take her home. They said they would wait for us downstairs while we got dressed.

The detective and the customs agent were standing at the front desk when we finally got downstairs. They wanted me to go straight back to Canada. Since I had paid for the room the night before, I was free to go.

Going outside to the parking lot, I saw my car wasn't there. I came back into the hotel and said to the detective, "Someone has stolen my car, because it's not in the parking lot."

They both looked at me and the detective said, "You parked your car between the two buildings. It can't be seen from the street or parking lot. It's still there."

I considered myself damn lucky for parking where I did, because had I not, the husband would have found us. It saved me from getting shot.

But why did I park there? The only reason I could think of was that the parking lot must have been full and it was the only safe place I could find to put the car.

The detective said, "The reason we know the car is there is because it is parked in the hotel owner's parking spot."

I retrieved my car and headed straight back to Canada. I kept telling myself how lucky I was that I had found that hidden parking spot and squeezed my Cadillac into it.

I relate this adventure just to illustrate how life is a game of luck. Even the border patrol couldn't find my car. Darlene's husband was looking for a white Cadillac whose owner's name was Peter, but he couldn't find it. No one knew it was there until the hotel owner told the detective that a white Cadillac was parked in his parking spot all night. You couldn't see my car even if you drove around the whole parking lot.

The border patrol officer knew my name because he was familiar with my boating accident on Belle Isle.

Dangerous Husband

There was one other episode with an irate husband with a gun. They always started with drinking and a woman who belonged to someone else. This one was no exception.

As I sometimes did when traveling back and forth from Windsor to Wawa, I stopped in a hotel bar for a few drinks to help make the drive easier. One time there was a great looking woman there that I struck up a conversation with. She said she was having problems with her husband and the more we drank the more sympathetic I sounded.

I finally realized that I was too inebriated to drive anywhere so I got us a room. We continued to drink in the room but it wasn't long before her husband found us and was pounding on our door. I answered the door and there stood the husband with a gun. He pushed his way into the room and started swearing at the women as he waved the gun about.

He was not interested in anything we had to say. He was causing such a scene the management called the police who arrived promptly. With guns drawn, the police called for the man to drop his gun and he got so nervous he pulled the trigger and a bullet went into the ceiling. He was then wrestled to the floor and arrested.

There I stood with my room trashed, a bullet hole in the ceiling, police all around, an upset woman and the management looking to see who would be responsible for the damages. I was so drunk by this time that I needed to find a place to sleep it off.

The next morning I woke up by myself in a different room and wasn't sure how I got there. When I was settling up my bill with the manager, he informed me that the night before I kept insisting that I needed a new room that was clean and quiet so I could get some sleep. They finally got tired of listening to me and moved me to another room.

Memorable Episode with Ted and Me

One time, when Ted and I were in Wawa, we found a quartz vein with free gold in it. I wanted to blow it up to find the actual gold vein. Ted, however, wanted us to dig up what we could see, because there was a lot of gold showing just on the surface.

I convinced Ted that I would work with him in digging out the gold. I then talked him into going into town and getting us something to eat and drink.

When he was returning, Ted heard a loud *kaboom*. I had gone ahead and blown up the rock to get at the rich gold vein I thought was there. But, as Ted surmised, there was no gold vein. It was all on the surface.

Ted then told everyone who would listen that he knew where the richest trees in the world were located.

All we had to show for such a promising gold strike was the following picture and the rich trees where all the bits of gold I blew up had landed.

Examining gold vein

MORE ADVENTURES

During the mid-1960s, I was laid off from the railroad because of the conversion from steam engines to diesel locomotive power. When the diesels took over, the railroad was able to cut staff to about one-quarter of what it was previously. This resulted in big savings for the railroad, because the operation and servicing of the new diesel engines was minimal.

There were times that I thought, to break the monotony, I would like to try a job that was different from any of the other jobs I had done in the past.

Viscount Motor Hotel

A new eighteen-floor hotel called the Viscount Motor Hotel was being built on Ouellette Avenue in Windsor in 1965. I applied for and got the job as chief maintenance officer for the entire hotel. My job mainly consisted of handling the maintenance workers and helping the construction crews. At the time I started working there, the hotel was only partially built. Although the main portion of the hotel was open and running, there were numerous contractor crews still working.

The finished hotel would include several bars and an entertainment facility on the seventeenth floor that was to be called "Top of the Town." It would feature dancing with a live orchestra.

The Corchis family not only owned the hotel but was also in charge of the whole building project. Mrs. Corchis was a very good person to be working for.

During the time of my employment there, I was under the impression that a major portion of the hotel was being financed through a Catholic church in Romania. The priest in Romania, who had arranged the finances, came to Canada to oversee the project.

Also during this time, Dick Thrasher who was a member of the Canadian Parliament had his constituency office in the hotel. I hadn't seen the priest for some time when I heard that Dick had the priest's immigration papers withdrawn, and the Canadian government eventually deported the priest back to Romania. I never heard the details of that situation but as I said earlier, it was a good place to work.

A lot of the guests in the hotel had leftover liquor in their rooms. My instructions to the maids were that any surplus liquor found in the rooms had to be put in a special storage room that I had and kept locked. Having that liquor available resulted in my doing a lot of drinking on the job.

Most of the contractor crews were Italian, and at the beginning of the workday they would always come to me and ask if they could get a drink to help them get started on their jobs. I would tell them that it was okay and asked what they preferred. In the storage room, I had wine, beer, rye, vodka, gin—in fact, any kind of liquor they would ask for or preferred. That was one of the main reasons I got along so well with the contractors, and they would immediately take care of any jobs or problems that we had to finish in order to keep the hotel running smoothly.

A young maintenance worker I had was named Carl, and he was an excellent worker and deserved a helping hand. I talked to the owner of McLeod Plumbing and asked him if he would give young Carl an apprenticeship to learn the trade as a plumber. Mac McLeod told me that he didn't have the authority to hire an apprentice. He had to get permission from the plumbers' union before Carl could start working as an apprentice with McLeod Plumbing.

I took Carl to the Plumbers and Pipe Fitters Union in Windsor, and I told them that Carl already had a job offer with McLeod Plumbing and that I would like the union to give him permission to start work as an apprentice. The head of the union issued an emphatic *no*—there was no opening available for an apprentice at that time.

It turned out that the Plumbers and Pipe Fitters Union ran a closed corporation and that the contractors could not hire any extra skilled tradesmen or apprentices without their permission. I was very surprised to find out that the union had so much control over the workers being hired by the general contractors and that the union ran such a closed operation in order to keep control of their workers, wages, and working conditions.

In the end, I had Carl signed up as a maintenance worker with the hotel. The contract was for a lifetime job as long as his performance on the job was satisfactory for everyone concerned.

I kept that job at the Viscount for approximately two years and really enjoyed it, because my wages were good and it had excellent working conditions.

Through Dick Thrasher, the son, George Corchis, had good connections with the government and received a government grant to open a tavern on Ouellette Avenue. Later, he was able to get another grant to start a tile company.

The Elmwood Casino Hotel was a well-known Windsor Landmark. The top performers in both the United States and Canada played in their wonderful nightclub, but it had fallen into great disrepair by 1979, when George Corchis bought it. Then, in 1983, he gave the deed to the property to the Brentwood Recovery Home with no payback due for the first two years. From November 1983 through July 1984, many unpaid people worked tirelessly on it and made it into a real recovery home. They also used the time to raise the money needed to own it outright. Brentwood Recovery Home still continues to operate today and has been a great success.

The Viscount Motor Hotel was sold in 1983, and the building was demolished in 1987. It was to be replaced by a condominium, but because of financial problems that did not happen.

Wringer Washing Machines

One of my partners from the treasure ship and diving school adventure was Ralph Smith. Ralph owned a large farm just outside Essex, Ontario. One day he was telling me about a storage problem he had with his barn.

He went on to say that a friend of his owned Beatty Washing Machine Company in Ingersoll, Ontario. Some time ago, the friend had gone bankrupt, and the company closed down. The friend asked Ralph if he could store his surplus Beatty wringer washers and parts in his barn.

Ralph asked me if I was interested in selling them so that he could get his barn cleaned out. I agreed, because I had quite a bit of spare time working at the Hotel.

I started by taking the wringer washers out of the barn and took them to my home in Windsor. The washers were very dirty, but they were all equipped with stainless steel tubs and had pumps that were a very popular make at that time.

When I started, I didn't realize that there would be such a demand for these machines. It was my understanding that the ladies used newer machines and there was no longer a market for these wringers.

The first six machines I brought home had to be cleaned because of the dust and dirt they had collected over the years. When I cleaned the stainless steel tubs and wiped off the wringers, they shone like brand-new. I put an ad in the Windsor paper saying, "Beatty wringer washer for sale—stainless steel tub with water pump to empty the washer."

I soon discovered that there was a terrific demand for them, and the phone kept ringing with people interested in buying one. The ladies wanted to use them to wash their husband's work clothes, dirty rags, etc.

A new washer at that time would cost between $500 and $700. I was asking $150 for these wringer washers. They sold like hotcakes. I kept working at them until Ralph's barn was empty. It seemed I was always busy with my car and trailer bringing wringer washer machines to my place, making repairs if necessary, and cleaning them in order to sell them.

I thought to myself at that time that I was really getting an education finding a hobby that was so lucrative. I feel that even to this day used wringer washers would have a big demand in the open market. Nothing cleaned work clothes better. It put extra cash money in my pocket, and it was enjoyable to do, because the person buying the machine would pay for it in cash and put it in the trunk of their car, and away it went.

It took me well over a year to run through all the machines in Ralph's barn. In the end I loaded the scrap pieces that were left over and took them to the scrap yard. Ralph was more than happy that his barn was now empty. I felt that I had given myself an education.

Hiram Walkers

Another time, I applied for a job at Hiram Walker Distillery in Windsor, Ontario, as the superintendent of maintenance. That job was going to be vacant within a matter of months because the present employee was going to retire.

They accepted me as a trainee to fill the job when the employee retired. They wanted me to come in and start working a few months early so that I could learn about the job and the work it required. It was basically repairing the machinery in the plant and running a machine shop for doing the repair work that was necessary from time to time.

I started to work at the plant, and at first it seemed like a nice, clean place that would be a good place to work. However, it appeared to me at that time that you could drink all the Canadian Club whiskey you wanted on the premises, but it was against the rules to take any bottles of liquor off the property.

Many employees, when they would report to work in the morning, would help themselves to some of the whiskey to help start their day. Then they would drink all during their working hours, providing they weren't drunk on the job. At noon hour, some employees had a habit of going over to the nearby Legion Hall, which they called "the Hut," for lunch, and they would be drinking beer over there.

It struck me that there was too much accessible liquor for a person of my type to be working at such a plant. I remember that there were two men there who had previously worked for me at the railroad. I especially remember one particular man named Sam who was had been a pilot in the air force during the war.

Sam would tell me about the hard time he was having because of all the drinking he was doing. Even though he worked full time, he said he was behind in his mortgage payments and didn't have the money needed to run the house and take care of his family.

That was another reason it was brought home to me that I shouldn't be working there either.

Also, during my time there, I drank heavily and came home drunk numerous times. After being in this condition for at least two weeks straight, I stumbled in the side door, and Marion and my daughter were at the kitchen sink doing the dishes. Marion had a cereal bowl in her hand, and as I came up the stairs, she pushed the dish into my face and broke it. Both Marion and I were dumbstruck about what had just happened. Marion calmly went to the telephone, called the doctor, and told him what she had done. The doctor just said to bring me into the office and he would fix me up.

This was the only time anything like this had ever happened, so I knew Marion was at the end of her rope in putting up with me. The time had come to quit that job.

Bob Dearth, a man I knew through my other activities, was the vice president of Ross Roy Advertising Agency in Detroit, Michigan. Less than two months after starting to work at Hiram Walker's, I phoned Bob to say I was short of money and required an investor.

Bob said, "How much money are you talking about, Peter?"

I told him, "I need at least $5,000."

Bob said, "Okay, $5,000 it is. If you come over to my office tomorrow at any time you can pick up a check."

I said, "Good, Bob. Good show. That's what I wanted to hear. I'll be there at 10:00 a.m. sharp."

That same afternoon at three, I went down to the office and told management that I was quitting immediately because I had other prospects—that $5,000 check from Bob Dearth—would get me started on another project.

I thought later that I had quit a good job, but because of the easy access to liquor, I had to get the hell out of there.

I later learned that Sam did go down the tubes due to his alcohol problems. If I had stayed at Hiram Walker's I probably would have been in the same shape as Sam, but I was fortunate enough to be able to get out of there when I did.

TROUBLED WATERS

Inventing a New Water Purifier

As I have outlined previously, over the years I experimented with magnets.

One day I was showing Ted Boomer my magnets and explained to him how they worked. When I was finished, Ted said, "Come on out and meet a friend of mine who sells water distillers."

I went with Ted and had a meeting with Guy Hamel, who sold water purifiers for Pure Water Society. I showed him samples of the magnets I had made for homes, and he said that they would be a wonderful item to use with his water distillers.

Pure Water Society's manufacturing plant and headquarters were located in Lincoln, Nebraska. The company invited the three of us out to Lincoln to view and inspect their facilities. I learned that the water distillers had problems with lime scale and hard water. We discussed the uses for magnets I had developed and I tested a couple of sets of my magnets on the water distillers, and they worked very well.

While in Lincoln, I was also introduced to Charles Thone, governor of Nebraska. During our meeting he asked about my various inventions, my prospecting for uranium and gold over the years and how my use of magnets could help in the water distilling process. The governor presented me with a certificate as an *Honorary Admiral in the Great Navy of the State of Nebraska*. This, he said, was in recognition of all my accomplishments. I never really learned who else these certificates were given to or the actual meaning behind them, but at the time it was fun to receive it.

After this trip, I understood the problems they were having with the water distillers. Guy said there was a big demand for a high-production water distiller that would be suitable for hotels, restaurants, taverns, and the like. Ice cubes made with distilled water were very clear and pure and seemed to be extra special when used in mixed drinks as well as in regular drinking water.

Ted, Guy, and I were talking about how I should come up with a high-production water distiller that would provide the volume of distilled water suitable for these businesses. That got me started thinking and working on developing such a system. I knew it was going to be expensive to experiment and build and rebuild different models, but I knew it could be done through trial and error.

We formed a large company, McLean Engineering, and began to raise money to carry the cost and give us some operating capital so we could do all the things that were necessary. I rented a store on Shepherd Avenue in Windsor and converted it into a workshop.

There were three of us on the board of directors: Ted was president, I was secretary, and Guy was a director of the company. We went to work selling shares to raise the capital we needed.

Guy Hamel had a large list of people who were interested in water distillers, and we found them very receptive to my idea of building a high-production model.

One of the conditions of the large distiller had to be that it would be cheaper to operate than a smaller distiller and have a large capacity. In reality, it had to be a completely new concept that we could patent.

I went to work on it with the idea of using the steam from a boiler to condense pure water. By using the feed water intake to the distiller to cool and condense the steam coming from the distiller, it would cut the cost of producing the distilled water down to one quarter of the cost of the machines that were on the market at that time.

The homeowners that already had Guy Hamel's distillers in their kitchen found them to be very expensive to operate because they were boiling water at full capacity. It is the same as boiling a kettle 24/7, which made it very expensive to operate.

The one I was working on had to be different. It had to produce more and cost less. I kept experimenting and improving a bigger unit in my workshop.

We always seemed to be short of money, and it kept Ted and Guy busy selling shares in our company.

Guy Hamel knew a group of people who lived in Edmonton, Alberta, who were very interested in our company. After discussing the matter, we decided that there was a large amount of money available out there if we could go to Edmonton and put on a presentation. We decided to go ahead and do it.

As I already mentioned, we were all short of money. Guy's credit card was maxed out, and my credit card and Ted's credit cards were heavily indebted. It was going to be a very expensive meeting. We had to pay for the airfare and accommodations and organize a dinner meeting at one of the upscale hotels in Edmonton. It would be a very important meeting—nearly life-and-death for our company—and we had to be successful in this money-raising project.

Ted was a complete showman and a terrific salesman. When we got there, he said, "My head isn't working the way it should, and I need a few double shots of vodka and water. In fact, I think I'll take some to the table with me."

I said, "My God, Ted, you can't be drinking before this meeting. There's too much at stake."

"Don't worry about a thing, and if I call on you or if you are asked any questions at the meeting, the most you say is just a few words. Make it very short, and do not interrupt me."

Ted had his shots and put the vodka on the table, and everyone enjoyed an excellent meal.

The presentation was about to start, and Ted said, "Ladies and gentlemen, I'm Ted Boomer and I'm very happy and pleased to see you people here. You're going to realize during my presentation that this is the opportunity of a lifetime. First I'd like to speak to Mr. [X] sitting right there in the audience. My understanding is that he is one of the most successful people and possibly the wealthiest person in the room. Now I'm going to talk directly to him. "I want you, Mr. [X], to interrupt me at any time to ask any questions that you think of."

Then Ted started talking about the benefits of our machine that we nearly had completed and gave them facts and figures of the cost per kilowatt gallon of producing the distilled water and the volume we were striving for.

Ted continued, "We're going to be producing, with this particular machine, four hundred gallons of distilled water per twenty-four hours. This means it's suitable for hotels, restaurants, taverns, and the like. In fact, our drinks tonight were made with distilled water. Everyone would love to have one."

He went on to say, "Naturally, this requires capital in order to finish the machine and get it ready for production. We are here tonight to raise capital through the sale of stock in our company."

Ted asked Mr. [X], "Do you have any questions? Is there anything further you would like to know? Is there any information I can give you that will help you understand the money and the increase in value of the shares that's going to take place in this company?"

The man said, "No, there isn't. I'm fully satisfied. I can see it is a good business opportunity, and I agree with all that has been said tonight."

"That's good to hear. How much money would you like to invest? And we want you to be the first to invest at this meeting."

"Ted, count me in for $10,000 worth of shares. I'll sign in with Mr. McLean, the secretary, immediately."

"Wonderful. Now I'd like to hear from the other people here."

The other people warmed right up to the occasion, and I had a line of people at the desk where I was sitting, accepting checks and writing out receipts. I was there about two hours, and the people put in various amounts ranging anywhere from $1,000 to $10,000. At nine in the evening we ended the meeting. Ted explained to the people that we had to return to Windsor and that our flight was leaving soon.

We put all the cash and checks away, gathered up all the papers, and called it a night. During that meeting, we went from being broke to having $195,000 in the kitty. We returned to Windsor and deposited the checks in the company's bank account, and I went back to working on the machine.

Peter James McLean

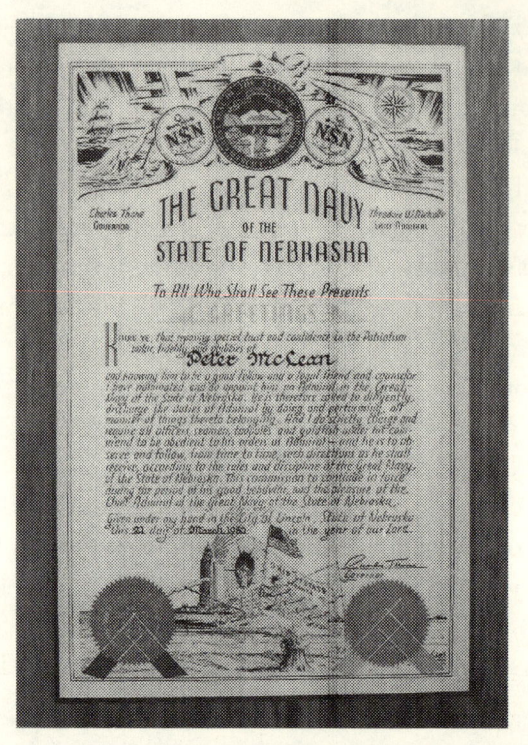

Certificate for honorary "Admiral" of the Great Navy of the State of Nebraska presented to me by Charles Thome, Governor of Nebraska

Heart Problems

During this time, I was having a medical problem but didn't see any need to visit the doctor even though I had previously had a heart attack. I ignored it until it was nearly too late.

I finally saw my family doctor, and he sent me to a heart surgeon named Dr. Asa. It was in the first part of January 1980 when I went into the hospital for surgery. Dr. Asa described the surgery as "giving [me] clean pipes." I never asked the doctor for any specifics about what he was going to do or why.

I was tied up in intensive care for four or five days and then in a regular room for a few more. I was worried about how they were going to take all the adhesive bandages off me. I actually got an award from the nurses for having the most stitches of anyone in the hospital at that time.

My family doctor had told me that Dr. Asa was the best I could have to do the surgery. However, I discovered Dr. Asa had the personality of a rattlesnake.

One day, Ted Boomer and Guy Hamel were visiting me, and they were sitting there smoking and asking how I was doing. I was telling them of my worry about how the doctor was going to remove the adhesive tape, because it covered my whole chest area and went down each leg. I told them the doctor would probably have me soak in a bathtub to loosen them.

At this point, Dr. Asa and a nurse walked into the room, and Dr. Asa yelled at Ted and Guy, "Get the hell out of here—you're smoking in front of a dead man."

Ted and Guy scrambled out of there in a hurry.

Dr. Asa said he wanted to look at me, pulled back the blankets, took hold of one strip at a time, and ripped the tape right off me. Talk about pain—it was everything I had worried about and more!

The doctor was very abrasive toward the nurse and told her to take out all the stitches. It took the nurse an hour to remove them all. It was not a pleasant time, but she tried her best to make it as pain-free as possible. When she finished, the floor looked like it was crawling with tiny black bugs.

Fortunately, I made a complete recovery.

Futures Market

It was also during this time that I developed an interest in the futures market and began to dabble in it. I bought and sold Japanese yen, silver, and sugar. I relate the following just to show how easy it is to win and lose in this market.

I remember when Bunker Hunt was trying to corner the silver market. The people in Washington didn't like what he was doing and charged him with having too many silver contracts. It wasn't illegal, but they stopped him from buying more contracts and put an end to his control of the silver market.

When the public learned what he was doing, the price of silver went from about five dollars to a high of fifty dollars an ounce. There was a lot of excitement and frenzy over the contracts being bought and sold. There was a real run on silver before the bubble broke.

Prior to this time, I had purchased four bars of silver that weighed one thousand ounces. I kept them on the floor in my den and no one ever noticed what they were. With the price of silver now around fifty dollars an ounce I decided to sell these silver bars and I made a hefty profit.

When I was admitted to the hospital and lying in bed awaiting surgery, I started thinking about the five sugar contracts I had on the futures market and thought I should sell them. I went to find a telephone, but the nurse wouldn't let me use it and said I had to stay in bed.

Here I was working on the water purifier, dabbling in the futures market and needing heart surgery. That was a load!

After the surgery, I was too sick to get to a phone for several days. I was tied up in intensive care for four or five days and then in a regular room for a few more. An even bigger worry than the price of sugar was, of course, how they were going to take all the adhesive bandages off me. I mentioned above how this was done and how right I was to worry.

It was well over a week before I could call my broker in London, Ontario, about selling the contracts. If the price of sugar had gone down, I would have lost my shirt, but instead of going down, it went up, and I made about $50,000. Go figure.

Four years later, I thought I might like to be a broker in the futures market. I had studied the market before investing any money in it and decided to write the TFE futures floor trader examination on December 13, 1984. When I was told that I had passed, I was asked if I wanted to become a trader and was offered assistance in the process of purchasing or leasing a seat on the exchange.

This was as far as it went. I just didn't have enough interest in pursuing this any further. Besides, my mind was on other adventures.

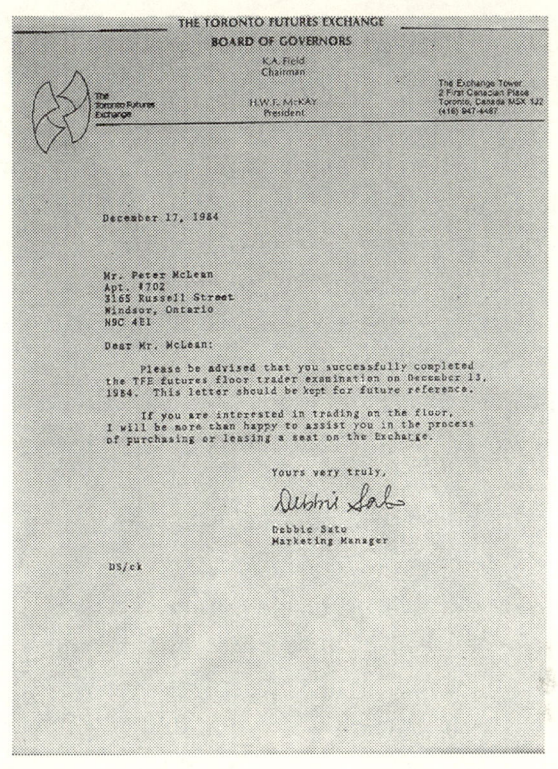

Permission for Trading as a Broker on the Toronto Future's Exchange

Back to the Water Purifier

We were starting to get worried, because the machine wasn't finished, and I was still a sick boy. When I got out of the hospital, I told Ted I had to have a couple of extra men to help around the shop, because there were things I couldn't do. The machine had to be built based on the plans in my mind, and I knew it could be done.

Ted sent the people in Edmonton information on the progress of the four hundred gallons per twenty-four-hour unit and told them that was what the machine was going to produce.

There was going to be a shareholders meeting in March 1980, which was less than three months away. We had the money, the material, and the help, and I got down to work. I got the machine working three days before the scheduled meeting. We tested it and it did four hundred and some odd gallons of distilled water in twenty-four hours. Ted, Guy, and I decided we were all set for the meeting.

Prior to the meeting, we had the shareholders come into the shop, and we showed them how the machine worked and how it was using the condensing feed out of the boiler. It meant we had boiling water going into the tank, and it wouldn't cost much to run the machine.

Ted, Guy, and I thought we would be heroes. We had our shareholders meeting at the Rendezvous Restaurant that night, and what happened was totally unexpected.

The main objective of the meeting was to elect a new board of directors. Ted was not elected to the board, I was not elected, and Guy was not elected.

There was a whole new slate of people elected to the board of directors for the company. The gentleman that Ted got to invest first in Edmonton was elected as a member of the board.

The shareholders had planned ahead of time and had their slate picked out for the company. The board of directors elected at that meeting wanted to dismantle the machine and take it to Leamington, Ontario where a new member of the board had a warehouse, and the machine would be under their ownership.

Can you imagine that between the three of us, we didn't have enough shares to control it? The founders of the company did not hold over fifty percent of the company shares. I can't begin to tell you how that happened, because I didn't understand it myself. But believe me—it did actually happen.

They bought out all of our shares at a good price, but we were no longer part of the company.

My workers and I took the machine apart piece by piece and left every piece there in a pile. I made sure that every piece of the new water purifier was there. The next day, three of the shareholders came and picked up all the material and took it to Leamington. They wanted me to go with them to rebuild the machine, but I refused.

The new shareholders couldn't find anyone who could put the machine back together.

In the end, the three of us were well compensated for all the hard work we had put into the company. The company, under new management, however, did nothing, and it just died.

Maybe I had known at that time but had forgotten over all these years that the patent for the water purifier was in my personal name and not in the company name of McLean Engineering. This meant that even if the new board of directors could get the machine back together and working, they did not hold the patent for it. The question of a patent for the water purifier was never raised.

Lawsuit

A couple of months later, Ted Boomer and I were charged with fraud. It was one of the American shareholders who laid the charge against Ted and me.

To fight these charges, it was necessary for us to hire a lawyer. George Yates, my friend and lawyer, wasn't available, and recommended we call Pat Ducharme, who was another well-known attorney here in Windsor. Mr. Ducharme assigned our case to an assistant but just prior to the trial the assistant was in a bad automobile accident and unable to continue. Mr. Ducharme took over the case himself.

Every morning we would meet in the coffee shop at the top of the courthouse and visit until it was time for court. The prosecuting attorney was Dave McIntyre. His neighbor had one of my early electromagnetic water purifiers hooked up in his cottage in Kingsville. Mr. McIntyre wanted me to install a set at his cottage because of the good reports that he was getting from his neighbor regarding how well it worked.

Mr. Ducharme didn't think it was right that I was sitting and having coffee with the prosecutor of my case. I told him that not once during our coffee sessions had we even mentioned the case against me, so there was nothing illegal about it. Mr. McIntyre agreed.

When we got to court and the case started, the American shareholder was the first witness. His attitude was that the machine did not produce four hundred gallons of pure water daily. Ted didn't take the stand, but I was called as the main witness for the defense.

When the prosecuting attorney was questioning me, I went into every detail of how I had proven to the shareholders at the shop the day before the meeting how the machine did produce the four hundred gallons per twenty-four hours. I explained in detail how we measured the water from the production of the machine with a stopwatch for timing the amount and then multiplied it out to establish the amount of water the machine would produce over twenty-four hours. The shareholders had been shown the machine operating and producing the four hundred gallons per twenty-four hours.

The American shareholder was again brought to the stand and was then asked if these facts and figures were correct. He stuttered and stammered and finally had to admit that they were correct and that he and the other shareholders had been present to verify the results.

The judge's verdict was that there was no case for any sort of fraud in regard to production of the water distiller. Therefore, the case was dismissed with costs going against the American shareholder who had laid the charges.

I told my daughter after the trial that if you are involved in a complicated case that included fraud charges, you should always call for trial by judge. Although you are considered innocent until proven guilty, it seemed to me that in a case of fraud, jurors feel that there must be some wrongdoing on the part of those being charged. That was why I chose trial by judge for this case.

Another reason was that the charges against us were such that Ted and I could have easily been sent to jail if found guilty. I have to admit I was quite anxious about the verdict. I told my daughter that I wasn't looking forward to being in jail. The verdict could go either way.

The judge ruled in our favor and we were found not guilty. Later that day my daughter said that Ted and I were acting like two schoolboys who had just gotten away with something.

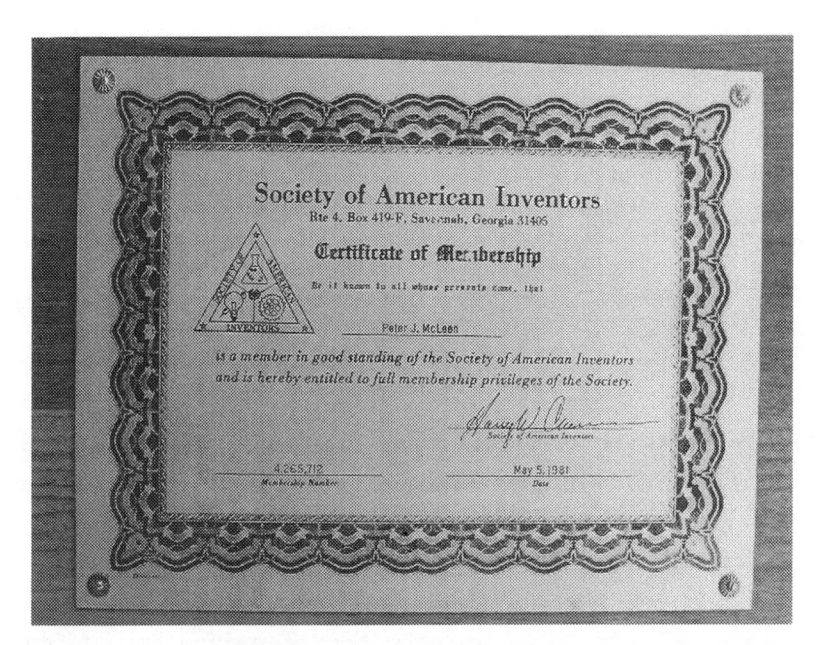

Society of American Inventors
Certificate of Membership

HOME AGAIN – ST. THOMAS

Railroad

During late 1960s or early 1970s, I was recalled by the railroad and had to move back to St. Thomas. I was still able to take time off, because again, there was always someone who wanted to fill in for me. Everybody wanted my higher pay.

To this day, the railroad companies say there is and has always been a no-drinking policy in force everywhere on the railroad. However, most of those I knew and worked with did drink, including me. One of the best ways to take care of inspectors and others was to empty part of a bottle/can of coke and fill it back up with liquor. If anyone knew, they ignored it. I did this often for others and for myself.

At one time Ted's brother was looking for a job on the railroad. I was able to help him get an interview, and he got the job. Since we didn't work together, no one knew about our connection. I was drinking heavily at the time and my associates on the railroad kept thinking I was going to be fired. The company did send someone to seek out the drinkers and give them notice. However, because the someone they sent was Ted's brother, he never even came near me. No one could understand why I wasn't targeted, and I never let on that I knew anything about it.

Diesel Engine Serviced by my crew
Myself in white shirt on Right

Marion

Marion and I had moved into a lovely brand-new apartment in St.Thomas, and she was very happy to be near her family again.

She decided, however, that she wanted to find a job. Because she was such an excellent cook, she applied for and got a job working in the kitchen at Memorial Hospital. She loved her job. The hospital had some long-term care patients who never had visitors. On her breaks and at lunchtime, Marion would visit with the different patients. Whenever our daughter and her young son were in St. Thomas, Marion insisted they all had to tour the hospital and visit with these patients. Needless to say, the patients loved to see her and whomever she brought with her.

Marion also spent about one year working in the kitchen at a psychiatric hospital in London, Ontario. However, she finally ended up at Elgin General Hospital in St. Thomas.

Peter James McLean

Toronto Adventure

While at Elgin General, Marion quickly made friends with the dietitian, and they became quite close. Because Marion was such a good worker and was interested in everything that went on, she took whatever courses were available. When she was asked to take the food supervisor's course in Toronto, she jumped at the chance.

Marion didn't really want to be in Toronto alone, so I said I would go with her. There wasn't room in her dormitory for me, however, so I stayed by myself in a motel room close by.

Before we left, I tried to think about what I would do with myself while Marion was at school, and when I finally came up with what I thought was a good plan, I brought my car, trailer, and tools with me to Toronto.

I contacted furniture and appliance dealers in Toronto and told them I would buy their traded-in washers and dryers that they had accumulated and unload them. The dealers liked the idea, and I was able to buy them cheaply for between ten and fifteen dollars each.

I went into an area known as Rosedale and located two garages that I could rent in order to clean up and repair the essentials on the appliances I had bought so I could resell them. Since the addresses of the garages were in such a high-class and respectable area of Toronto, they drew a good response when I ran ads in the newspapers.

All during the time I was in Toronto with Marion, this is what I did. It kept me busy during the day, and I was then able to take Marion out for dinner in the evening and just spend time with her.

At the end of our stay, I had two garages that were still half full of the appliances that I hadn't yet sold. I contacted used furniture dealers on Queen Street and gave them a sad story about how I had just broken up with my wife and had these appliances left over in storage and they needed to be sold immediately.

One of the dealers came down and looked at what I had in both garages, and I made a deal with him for cash for what was left. I also gave him the keys to the two garages and told him that they had to be emptied before the end of the month, because that was the end of the rental period.

What I did was a lot of fun, and at the end of my stay in Toronto, both garages were empty. It proved to me that if I was ever stuck in a strange town, I could make a living with a similar setup.

I have to be honest, though, and say that during our stay, I had so much free time on my hands that I easily got into trouble and I was also drinking. There were rumors that I actually sold appliances that didn't work or had parts missing. I was never charged with anything and felt, overall, that the trip was a great success.

In fact, I more than paid for our time and expenses in Toronto. It was a very pleasant time for Marion, and she did well in her course and graduated at the top of her class. When she got back to work, she was promoted to food supervisor.

During my time back in St. Thomas, I kept my mining claims and worked on them to keep them up-to-date. Also, there were a lot of other adventures that kept me busy.

Excelsior Truck Leasing

This company owned all the trucks and equipment that were used on the railroad. Because of tax write-off rules, they only held the trucks for a period of five years. After that, they were put up as surplus material and sold on a personal basis to people they knew who were interested in them.

I knew the manager of the company in Detroit very well. He and I got to be friends during the late '60s and early '70s. He was very interested in Canadian silver coins that had a high silver content. He wanted any that were prior to about 1962. I had jars of these coins that I had saved over the years, and because of his interest I would sort them out and keep him supplied with all the coins that he liked to have.

As a result, I had first choice of the trucks that were available to be sold. The trucks were very well built, because they had ten-ply tires, heavy-duty brakes, and equipment on them. When I got possession of the trucks, I didn't have to transfer the plates or ownership, and I could sell them on a personal basis without the expense of transferring the titles, etc.

Whenever I got the trucks, I would advertise them for sale after I had the boys in the diesel shop clean them up and repaint any portion of the vehicles that was necessary. I could sell the trucks to private individuals with very little effort or expense. I had good luck selling them, and everyone knew that I had these trucks available and would ask me ahead of time for a specific truck that I had coming up so they could be first in line to buy that truck. I was doing this for part-time income.

One truck that came up for sale was one of the railroad's heavy-duty work trucks with a boom attached, and it generated a lot of interest. We lived at that time in a new apartment building on the west side of St. Thomas. I had the truck parked across the street in a gas station.

About three o'clock on a Sunday afternoon, I was called to the lobby by a person who wanted to see me about the boom truck. When I got downstairs, there was a real rough-looking character waiting for me, and he introduced himself as Joe Carter and said he wanted to buy the boom truck because it was very important to him.

He said, "I pick up scrap and such from farmers and take them to a scrap yard in Hamilton, Ontario, and I could really use that truck."

The fellow reeked of liquor, and I told him, "You can't buy the truck without money, because a lot of people want it. I need cash for it."

He pulled a big roll of money out of his pocket and said, "I can give you most of the money right now."

"I don't accept a deposit, but when I transfer the truck to you, I have to have the total amount of money in cash."

"I've got to go to the bank first thing in the morning, and then I'll meet you and pay you in full."

I was emphatic when I told him, "You have to come before noon with the cash, because after that, the truck will be sold to someone else."

The next morning about ten thirty, he came to my office at the railroad and said he had the cash with him. I said okay, and we went to the license bureau and transferred the title over to him. It was a $9,500 cash deal. He was such a rough-looking character that I didn't think he had the money, but he surprised me, and the truck was transferred and gone.

Confessions of an Eccentric Dreamer

About two weeks later, I had been down to Niagara Falls, and when I was coming back, I stopped in a hotel on the highway in Tilbury to have a drink. Joe was sitting there drinking beer with a couple of friends.

He said, "Hey, Pete, I'm glad to see you. Things are really going along good with the boom truck. We just got back from Hamilton. On the way there yesterday, we stopped in Simcoe and scooped up a late-model Corvette with the boom and stripped the plates from it, upset it into my trailer and took it to the scrap yard, where the boys stripped all of the valuable parts that they wanted to save and then put the car through the crusher."

He continued, "I really appreciated you selling me the boom truck, because I'm making good money with it."

I told him, "Okay, I just hope you don't end up in the slammer. Good luck to you," and then I left.

At other times, Joe was after me to buy other trucks I had available, but I kept telling him that they were already sold. I used that as an excuse so I didn't have to sell any more trucks to him. I didn't want to know what he would use them for.

Sorting Coins for Excelsior Truck Leasing Manager

THE ITCH IS BACK

Once again I was getting restless in St. Thomas, and I decided that I wanted to return to Windsor. Marion finally agreed to move with me, and she easily got a job as a food supervisor at Grace Hospital.

At first, we bought a small house, and I again turned the garage into a working den, where I did my research looking for more mining property to develop.

This worked out for a time, but I soon started drinking heavily again.

At one point, my oldest daughter, Barbara, had to call an ambulance to take me to the hospital to get dried out, but I refused to go. The EMS cannot take someone involuntarily out of his or her house. Barbara told me later that when they had me strapped on the gurney, she tucked a bottle of liquor in beside me so I would relent and let them take me to the hospital.

I finally got straightened up again. However, after a time Marion and I decided we would do better living in an apartment. So I packed up all my books, mining supplies, ore samples, etc., and put everything in storage.

We looked for and finally found an apartment in the new Marine Court Apartments on the Detroit River near Sandwich and Mill Streets. The only problem we had there was that Marion had to leave at five in the morning to get to work on time, and she didn't like leaving that early. I told her that if we found another apartment closer to the hospital, I would walk with her to and from work.

I started looking for another apartment, but everything within walking distance to the hospital was full. However, I met a man by the name of Ivan Vexler, who owned several apartment buildings as well as a clothing and shoe store on Ottawa Street here in Windsor. He was a very nice fellow, and he gave me the job of managing one of his apartment buildings that was located within three blocks of the hospital. Marion could now walk to work, and as promised, I walked with her and then would meet her after work and walk home with her.

Managing the six-story apartment building was very enjoyable. The carpets, the halls, and the stairwells were my biggest job. They had to be cleaned and vacuumed regularly. It was the only job I absolutely hated, and I tried to come up with a way to get someone else to do it.

I used my personality plus on the tenants and took good care of them and their problems. In return, the tenants started to vacuum the hallways adjacent to their apartments, including the stairs and stairwells. This made my job as the apartment manager very easy and enjoyable.

Marion Retires

Marion and I lived in Windsor until she retired from Grace Hospital in 1985.

Our youngest son, Peter Jesse, had just separated from his wife and had custody of his newborn son. Marion wanted to be in St. Thomas so she could look after the baby, and she bought a house there. Notice I said that *Marion* bought a house.

Needless to say, I was not particularly happy about moving back to St. Thomas since I loved being in Windsor, but I finally did make the move. Although we were again in St. Thomas, it didn't interfere with any of my adventures. I continued prospecting and developing mining properties.

However, at one time or another, I would decide I wanted to work on another adventure that I felt I couldn't do in St. Thomas, so I would just up and leave. I would take all my belongings, personal and otherwise, and move to Windsor by myself. I did this at least two or three times.

Each time, I would rent a house and try to conduct business, but, as I always did without Marion, I would self-destruct each and every time. My drinking would take over my life. If I'm honest, I have to say the houses I rented were lovely places, but I did not keep them up. They probably required a lot of cleaning after I left. One rental house even came with a wonderful dog who loved it when I took him out in the car to get an ice cream cone.

When I'd finally realize I needed stability to get straightened out, I'd pack everything up again and return to St. Thomas. Marion would just say, "Oh, you're back again" and take me back every time. She did make it very clear to me, though, that I was a "boarder with benefits," because she owned the home in St. Thomas. I paid $500 a month to live there.

Many, many years later, I finally squared up the finances with the house and Marion, and at that point we owned the property jointly.

Marion on the porch of her new home in St. Thomas

Income Tax Woes

For prospectors, Revenue Canada is very good as far as paying taxes are concerned on any money they make. For instance, if prospectors have staked claims and sold them, then that money is not taxable. It is tax-free.

However, other monies that they would make through labor, working, etc., are all taxable. As I mentioned previously, the majority of mining executives call themselves executives when in reality they take advantage of the tax allowances that are beneficial to prospectors only.

I had a tax problem with Revenue Canada for quite a few years. The problem was that even though I called myself a prospector I wasn't classified as a full-time prospector because I was employed by the railroad. In reality, therefore, my full-time employment was with the railroad.

Revenue Canada, in approximately 1980, presented me with a statement that showed I had a large amount of income tax owing.

I had a small bank account at the Bank of Nova Scotia in a strip mall on Huron Church Road next door to a library branch here in Windsor. The bank called me one day and said the Revenue Canada had seized the money in this particular bank account; the amount was between $1,100 and $1,200. That spurred me into action.

I asked my friend and lawyer, George Yates, for the name of the best tax lawyer in Windsor, and he referred me to R. Bruck Easton. I made an appointment with Mr. Easton to discuss my tax problems.

After reviewing the papers I brought him, Bruck's statement was "Peter, you owe the tax department a lot of money, and they are certainly going to make every effort on their part to collect it. For us to fight the amount and argue about the taxable income is a very expensive and lengthy proposition."

So I said, "Bruck, this is what I want you to do. I would like you to contact the tax department and tell them, on my behalf, that I plan on paying these taxes owing as soon as possible. I have been and am currently doing prospecting full time since my retirement from the railroad. And it is possible that I could strike it rich at any time or at some time in the near future. I would like you to stall the collection of any taxes owing for as long as we can."

Bruck answered, "That's the best idea I have heard. Now I'm going to use my influence with the tax department to postpone the paying of your taxes as long as we can."

Bruck contacted them through his connections with the tax department, and they said it made sense for them to hold off the collection of taxes until further notice. However, the interest on the income taxes owed would carry on and be applied to the total amount until paid.

I would contact Bruck about once a month to ask how much I owed the tax department and ask him to keep postponing any collection, which he did. The legal cost for Bruck was not very much for his part in the postponements. However, his costs would increase substantially as the case moved forward through the court.

This went on for years. In 1985, I had moved from Windsor back to St. Thomas when Marion retired. One day Bruck called me and said that he had postponed the payment of the taxes as long as possible.

By then, I had retired from the railroad and was getting a pension. Bruck said the tax department was planning to take me to court in the near future in order to establish a collection program for the monies owed. I told Bruck I'd be up to see him soon. I asked him for a rough estimate of how much the fee was going to be to handle the case through the courts.

On the day I went to see Bruck, I said, "I didn't want to tell you the bad news any sooner than I had to. However, Bruck, I've been retired from the railroad for a while now, and I've had some bad luck financially with projects and prospecting which I've been doing." I then said, "I'm unable to pay you the fee I owe you or the fee to handle the whole case."

Bruck nearly fell off his chair. He said, "My god, Peter, I'm involved in a firm here in this office, and I have to collect my fees to go into the firm's account."

"Well," I said, "the only way out is for me to get legal aid. That way you will be paid your money through them as well as the cost of the court case."

We discussed the matter and went back and forth until I finally convinced Bruck that legal aid was the only possible way out.

He said, "Have you applied for legal aid yet?"

"No."

"Well," he said, "you're living in St. Thomas now, and legal aid has to be applied for there, because that is where your residence is."

"Now," he said, "what I'm going to do is contact the legal aid people in St. Thomas and explain your circumstances, and I'm quite sure I'll be able to get them to help you. I'm going to write them a letter, and I want you to make an appointment to go in personally to meet with them."

About a week later I took the requested paperwork and went to the courthouse in St. Thomas to the legal aid department. I had a meeting scheduled with the person in charge to explain my circumstances.

When I got there and told them who I was, the lady in charge said, "Yes, Mr. McLean, I've got a letter here from your lawyer, Mr. Easton, in Windsor concerning the legal aid help that you require. After reviewing everything, I'm approving you to receive legal aid to cover the costs of this case."

Incidentally, she showed me the letter that Bruck had written, and as I read it, I realized that it was the nicest, most well-written letter I'd ever recalled seeing. To me, the letter was a masterpiece.

The amount of the tax bill at this time was approximately $350,000, including all the interest that had been accumulating.

I phoned Bruck and said, "Legal aid approved me, so now we are all set to go."

Bruck replied, "I'll notify you as soon as I know when the court case is scheduled. It will be held in London, Ontario and I'll be there with you."

He added, "Peter, you have to make sure that nothing gets in the way of you being in the court with me at the time and date specified."

A couple of months later, I heard from Bruck, and he said, "The date is set in London. I'll be there, and we should meet in the food court at the downtown mall an hour before the court case. Then we can proceed to court together."

I met with Bruck as planned, and we had coffee and donuts and discussed things in general, because he had said, "We can't discuss the court case because we don't know what's going to happen, but when we get into court, I'm going to do the best I can for you."

We went into the court at the proper time, and the judge sitting there was the most shabby, hard-looking character I'd ever seen in a courtroom. He was unshaven and looked like he had a mammoth hangover.

Bruck started with, "Your honor, I'm not sure how we can proceed with this matter, but I certainly appreciate your handling the case, and I would like us to work things out to everybody's satisfaction."

The judge said, "Well, thank you for coming, Mr. Easton, and I appreciate your being in court, but I'm looking at this case, and what we've got is one government department in conflict with another department."

He continued, "We've got Revenue Canada on one side and legal aid on the other. To my way of thinking, it's government against government, and we have to come to some monetary settlement here and now. I want you to take your client outside the courtroom and discuss with him the largest amount of money that he can come up with to pay this court, and the matter will be settled."

Bruck said, "Come on, Peter, let's go outside the courtroom."

We went into the men's washroom. Bruck said, "Now, Peter, you've got to come up with enough money here and now to settle this up. The amount of money I'm thinking about is at least enough to pay for the expenses of the court case. If you can, I'm pretty sure that we can get this settled here and now. Now, Peter, how much money can you come up with?"

"Bruck, you know I'm hard-pressed for finances, but I can pay $2,500 right here and now to the court."

"Is that all you can come up with?"

"That's all I have."

"Okay, let's go back in."

Bruck presented to the judge that his client could only come up with $2,500 cash to pay into the court this morning to settle this case.

The judge slammed his gavel and said, "That amount is acceptable. Case closed, settled, and dismissed."

He finished with "Thank you, Mr. Easton, for your help and cooperation in this difficult matter."

I wrote a check out for $2,500, which is all I had in the bank, presented it to the court secretary, and got a receipt, and the matter was settled.

Just as a game of luck or good fortune, the only thing that saved me in court was having legal aid. If I'd been there paying out of my own pocket, I would have ended up paying the total bill on some kind of a monthly payment plan. But it ended up with legal aid paying Bruck the money that was owed to him. The judge realized that a person on legal aid was not going to be able to come up with any future payments.

When Bruck and I got back to the food court and were drinking coffee, Bruck said, "Son of a bitch, I'll be damned."

He sputtered and said "it's the damnedest court case I have ever been involved with." I just can't get over the fact the case has been settled in the men's restroom in a courthouse."

Bruck and I ended up being good friends, because he realized things couldn't have worked out better for him; his firm in Windsor got paid what it was owed, and the case was settled.

Currie Rose Resources Inc.

After the settlement with my Golden Goose Gold Mine and Deep Lake Gold Mine, I was financially secure and again had my freedom.

I still had gold fever and staked a lot of mining claims in Lundrum Township and Rabazo Township.

I also staked claims surrounding the Wawa Creek Falls area. I filed claims here because it was such a beautiful area and I always felt sure that there would be a large deposit of gold accumulated at the bottom of the falls. I also owned claims on the Dead River portion of the Michipicoten River.

During my prospecting and panning for gold, I was taking samples from different locations. I would bring the samples home and extract the gold in the quiet of my backyard.

I had bought a diamond tester from Bob Rose, a friend and jeweler in Rochester Hills, Michigan. The tester would positively identify a diamond—if you were lucky enough to find one to test. Diamonds are very hard to identify because they don't shine and have a very dull appearance. I had tested between thirty and thirty-five locations, but I didn't keep tract of these locations as well as I should have. I tested all the rocks and chips with my diamond tester.

I told Mickey Clement and showed him some diamonds that I had found, and he wanted to know where I found them. Since I had tested in so many different areas, I couldn't tell him exactly where they came from.

Mickey knew the chief government geologist of the area, and he was going to send the chips to him to be tested. One turned out to be a three- or four-karat diamond but it was only industrial grade because of the color and fractures.

In the *Northern Miner* magazine at a later date, Mickey had a small write-up that said an old-time prospector in the Wawa area had found a couple of diamonds but wasn't sure of the exact location where they were found. This was the beginning of the diamond rush in the Wawa area. There were about twelve to fifteen different companies that began to explore for diamonds in the Wawa area. At that time I had approximately thirty-five or forty claims staked and registered in my name in the area.

One of the companies that came to Wawa was Currie Rose Resources. One night while in the bar at the Wawa Motor Hotel, I was approached by a fellow named Harold Smith who said his company was a newcomer to the area and wanted to buy all the claims that I had available. Harold Smith was the president of Currie Rose, and he had the secretary/treasurer of the company with him. We all sat down and went over the maps showing the locations of the various claims I had.

We made a deal for all my claims in Lundrum and Rabazo Township for one hundred thousand shares of Currie Rose stock and $100,000 in cash.

The next day, we went to the Department of Natural Resources, and I transferred all the claims that were listed in my name to Currie Rose Resource Inc. Their shares were listed on the Vancouver Stock Exchange, which meant that a broker could sell the shares at any time.

Currie Rose was well financed, and I kept thinking I had to hang on to these shares because of the possibility of the company finding a large diamond deposit in relation to the property I had just sold to them. If that happened, the shares would skyrocket in price. I held onto the shares, and I still have them today. The company spent a lot of money, but it didn't turn out to be successful.

After this, I didn't have any claims left in the Wawa area. I was cleaned right out. However, I now had the finances to proceed with prospecting and staking further claims.

Permanent Prospector's Licence

Mishibishu Lake

After the deal with Currie Rose, as I said before, I had no claims and no prospects left to work on.

One day, Mickey Clement and I were flying to Marathon, Ontario, and we went over a lake about a half hour flying time out of Wawa. The lake was striking, because it had a lovely bay and beach on one portion of it. Also, there were a couple of old lumber shacks built along the shoreline with a long wooden dock. I asked Mickey what the name of the lake was, but he didn't know.

When we got back to Wawa, I went into the Department of Mines and checked over maps and such until I found out that it was Lake Mishibishu. The property had gold showings on it. Previously, some company had been working on the property, and it turned out they were the people who built the cook shack, bunkhouse, and dock.

I got very interested in the property, and when I investigated further, I found it was open for staking. I immediately staked six claims along the bay in order to get things started.

There was no road to the property, so it was necessary for me to go back and forth by seaplane. I was using White River Air Service in Wawa, which had an old Beaver aircraft with no seats but lots of room for storage, supplies, and four or five people. I flew in with drills, picks, shovels, food for the cook shack, dynamite, and all the other equipment I would need to start prospecting on the property.

George Kenty, my friend and a mining engineer, helped to direct the workers I had flown in to the property. We established pits about one hundred feet apart and the overburden was cleared. We found pockets of high-grade gold. George recommended that I stake at least another eighteen claims in order for me to protect the property and to include any further gold bearing ore we were likely to find.

I was working on the property out of my own pocket, and it was getting very expensive having to fly back and forth with the supplies, equipment, and workers. It required more money than I was able to afford out of my own pocket. A lot of the high-grade ore samples that I have in my collection came from this property. The different pockets in the quartz veins that had high-grade gold showing in them were tremendous and very exciting.

I kept thinking to myself that with the right people, this property would be an easy sell.

I ran ads in the *Detroit News*, the *New York Times*, and the *Globe and Mail*. The response was practically nil. Finally I had read about a newspaper called the *Journal of Commerce* (a New York newspaper) and I ran an ad for investors or buyers for high-grade gold property in Northern Ontario.

From that ad I got a response from a mining engineer from the New York area who was very interested in the property and wanted to hear the details from me. We talked by telephone, and he agreed to bring two of his partners and fly up to Wawa to visit the property. I was to pick them up at the airport at the agreed date and time. They arrived at approximately eight in the morning on a private jet.

I told the White River Air Service that I had planned on taking three visitors out to the property on that date. The day turned out to be bright and sunny. I brought a couple of cases of beer and ordered a big takeout lunch from the Wawa Motor Hotel. I wanted these gentlemen to have enough food and drink while at the property.

We flew in, and I started a fire in the cook shack in case they wanted coffee. I explained to the mining engineer how the property was laid out and that I had opened the gold vein with the different picks and traced the vein for over one thousand feet.

We were on the dock, and I had a wooden container under the water that I used to the keep the beer cold. When you store beer in such a situation, it has to be covered, because if the sunlight hits the bottles, it will ruin the beer.

We were talking, and I asked them if they cared for a sandwich and coffee. They said they just wanted to stand there on the dock and walk around the beach in the sunshine. The engineer was not interested in examining all the different pits and trenches that I had opened with the gold showings in them.

As we were standing on the dock looking over at the horizon, suddenly, at treetop level, an old DC9 aircraft equipped with bomb bay doors was flying right at us, coming in very low. The plane opened the bomb bay doors and dropped a load of young salmon.

The three men got all excited watching this and wanted to know the story behind the drop. I told them that the Department of Natural Resources was restocking the lake. The salmon would be going from Lake Mishibishu down some tributaries into Lake Superior. I also told them there was no fishing allowed, except by the people who owned property on the lake.

They indicated they were very interested in the property, but I still couldn't get them interested in exploring it.

After spending about three hours roaming around the dock and water's edge, they said they were not interested in mining. They had something different in mind for the property. They asked me to get the seaplane to pick us up so we could go back to Wawa. They had not eaten any of the food I had provided, but they did love drinking our Canadian beer. I was very discouraged and felt they were only on a sightseeing tour.

Raising a flag on a pole was the signal to tell the seaplane to pick us up. The pilot saw the flag, and about thirty minutes later, he landed and flew us back to Wawa.

At their request, I took them to meet the manager at the Department of Natural Resources. We went into the manager's office, and I told him that these three gentlemen were from New York and were interested in my property on Lake Mishibishu.

"These goddamned Americans are always trying to buy something," he said. He got very hostile and yelled, "You sons of bitches! The lake is out of bounds completely for people who don't own property there."

He was very abrupt and had a real hatred of Americans. However, the Americans just appeared to ignore him and everything he said.

When we left there, I said, "Come on down to the Department of Lands and Fisheries, where you can get the full story explained to you about the seeding of the salmon, fishing, and property ownership."

I knew the manager, who was a gentleman by the name of Jim, and I introduced him to the Americans. I told him, "My guests are interested in buying my mining property on Lake Mishibishu, and they would like to know all about reseeding the salmon and fishing restrictions and different laws and regulations in connection with the lake."

Jim was glad to help and gave us some government publications with all the facts, conditions, laws, and regulations in regard to the lake. He wanted us to take the publications back to the motel with us and study them at our leisure.

We went back to my room at the Wawa Motor Hotel, where I showed them the maps of the property that I had staked and owned. These claims included all of the bay area, docks, buildings, etc. They read the publications about the restrictions on the lake and hinted to me that they had been looking for property such as this for a long time. They asked, "If we buy your claims, how long would it take for you to transfer ownership?"

"I'd have to get you a miner's license to transfer ownership of the property. I can get the miner's license and sign the transfer papers right in the Department of Mines."

"We want to buy the property lock, stock, and barrel."

"Hold on," I said. "This is not going to be a cheap transaction, because the property has to be paid for in cash."

They wanted to know what I had in mind, and I told them, "With the amount of effort and expenses, the property is worth at least $250,000."

They were flying in an executive jet, so I was thinking to myself that these gentlemen certainly had money.

"We can't go that high. We aren't interested in mining. Our only interest is in ownership of the property, because we have other plans for it."

We settled the deal for $200,000 cash.

When I asked them about shares in the company along with the $200,000 cash, they said, "No, no shares, only $200,000 cash tomorrow at the bank. We'll stay the night in Wawa, and first thing tomorrow morning, we'll complete the deal and finish the transactions. Right after that we'll be flying back to New York."

The next morning we went to the bank and they secured a bank draft, which is as good as or better than a certified check. We then proceeded to the Department of Mines and completed the transfer papers.

Everything was signed and sealed over to them, in the name of a company I had never heard of. Then, I had my money transferred to my friend and bank manager, Garvey Shearon, at the Bank of Montreal in East York, Toronto.

I couldn't believe what had just happened. The only thing I could think of was that these gentlemen were going to use the property for pleasure, not mining. It was one of many strange things that have happened to me.

My "Den" and workshop in St. Thomas

MARION AND ME IN LATER YEARS

When our boys were young, I had a friend, George Abar, who owned a tavern on the east side of Windsor. Every Christmas Eve, George would dress up as Santa Claus, come to the house late at night, and bring the boys their toys. It was a fun time for everyone, but it wasn't long before the boys out grew Santa Claus.

Marion and I then decided we wanted to spend Christmas in Florida where it was warm. We looked forward to our trips there every year, and since we stayed at the same motel in Fort Lauderdale, we got really close to the owners. We went there for many, many years.

After Marion retired to take care of our grandson Shaun, she didn't want to travel to Florida anymore, so for several years, along with my two daughters, Barbara and Carolyn, I would drive to Orlando for a couple of weeks of fun in the sun. We always enjoyed our time together. To make sure the girls were independent while we were there, I always rented a car for them to use.

Around 1990, Marion was diagnosed with macular degeneration. It wasn't severe in the beginning, but she could no longer drive and had to sell her car. She was not happy about it, because she loved her car and the freedom it provided. Shortly after this, she was diagnosed with dementia, but that too was not severe in the beginning. However, the two problems gradually began to worsen.

In about 1996, Marion decided she wanted to fly down and meet the girls and me in Florida and bring Shaun with her so we could have a family vacation.

The day we were to pick them up at the airport in Orlando, we got a phone call that Marion had had a mild stroke and was in the hospital. I immediately called her, and she was adamant that we not return home. She had already arranged for Shaun and his mother, Patty, to fly down to Orlando to meet us. Marion's two sisters took turns staying with her until we got home. We talked to her by phone at least twice a day.

After we returned home, it didn't appear that Marion's stroke had left any residual side effects, but her eyesight and dementia were slowly getting worse.

There was a dinner at a rhubarb festival in a nearby town and I thought Marion needed to get out of the house so I took her. Between getting out of the car and sitting down in the hall for dinner, Marion's demeanor suddenly changed and we soon learned she had had another small stroke. It was amazing how it could happen so quickly. It also created more problems with her dementia.

Marion began to accuse us of moving her to a new house every day. She would get up, get dressed and then ask to be taken home. We would tell her she was already home but she would insist that she had again been moved.

Another time as Marion and I were getting out of the car, Marion fell and hit her face. Fortunately our neighbor, who was a paramedic, saw what happened and helped me get her to the hospital. By the time the girls got to the hospital from Windsor, Marion looked like a very battered woman. They released Marion the next day, and the girls went back to Windsor a day later.

A couple of nights after she got home, I could hear voices in the living room and went out to see what was going on. There, I found two female police officers talking to Marion. She had called 911 and told them she was being held against her will. She wanted them to take her home to 60 Lois Avenue. That was exactly where she was.

I explained about the dementia and that she was constantly asking to go home. Every day she accused me of moving her to a new house. We lived in a one-floor ranch home, but she told the police she was being kept on the second floor and couldn't find the stairway to get down to the front door.

As they were leaving, the police officers recommended that we look into moving her to a nursing home. When I told the girls about the visit from the police, they kept thinking I could have easily been arrested for spousal abuse because of the bruising on Marion's face from her fall along with her saying she was being held against her will. Thankfully the police understood the situation.

Along with Barbara and Carolyn, I finally realized that Marion needed to be in a nursing home, so we went through all the necessary paperwork and had her booked into one not far from us. It was very hard since she couldn't see and her imagination was working overtime. I tried to have lunch with her every day and I often took her out for a drive. She loved to go out for an ice cream cone. Still, she was so unhappy that I started to bring her home for weekends. Barbara and Carolyn would drive down to St. Thomas to help. This went on for a very long time.

Marion also had to have surgery in London, Ontario, to have a shunt inserted to help relieve normal pressure hydrocephalus. This condition mostly affects people over the age of sixty, and it was felt it could be adding to her dementia. The doctors hoped the surgery would help, and it did help somewhat.

In July 2004, Barbara, who was a registered nurse at Henry Ford Hospital in Detroit, Michigan, turned 65. On one of the girls' weekend visits, Barbara decided to retire and move to St. Thomas to look after Marion at home. This made it possible for Marion to come home to live. She still thought we kept moving her every day from one house to another, but on the whole she did quite well for a time.

It was wonderful having Barbara with us, and it left me time to work with Carolyn in the writing of this book. It turned out to be a much larger project than I had first envisioned. Also, it took much longer than I'd thought it would. Carolyn still lived in Windsor with her husband and couldn't come to St. Thomas for more than a few days at a time. She refused to leave her husband and move here permanently. I couldn't figure out why.

There came a time I was sitting in my office and dreaming of another adventure since there were still a few things I wanted to do but had never had the chance to do before. I had been giving serious consideration to leaving Marion and moving back to Windsor to pursue another of my own dreams. I clearly remembered how I would self-destruct whenever I was on my own but still thought I should take a chance and do it.

All of a sudden, I heard a loud voice say, "Peter, you are not going to leave your wife. You have to stay here and take care of her!"

I was so startled I nearly lost my breath. Where the voice came from I do not know, but I did know that it meant business. I spent the next few days thinking over my options and decided the voice was right; I had to stay home and take care of Marion. I never regretted my decision. My girls said it was my Mother who told me I had to stay because she truly believed that when you married you stayed with your spouse forever, regardless of the circumstances.

In late 2006, Marion's condition deteriorated more quickly until finally Barbara realized she could no longer take care of her at home by herself. Marion needed to be in a nursing home. We again went through the paperwork and had Marion booked into a brand-new nursing home by the first week of January, 2007. On a Friday we got her all settled in her new living quarters.

The next day, Barbara was driving Carolyn back to Windsor, and they stopped to see Marion on the way. She was picking at her breakfast and kept telling them, "If you leave me here I'm going to die". They didn't leave until she was again settled in her room and in bed.

Also on that day, Marion wanted me to have lunch with her, so I went out and joined her. She didn't seem bad to me. However, about two hours after I got home, I got a call from the nursing home telling me that they couldn't let me come back to the home without getting permission from the staff. It turned out the Norwalk flu was spreading fast throughout the facility. When Barbara got back home, we both went to see Marion, but again, they wouldn't allow visitors.

That same night we got a call that Marion had been transferred to the hospital in St. Thomas. Barbara and I went immediately to the hospital, but her body was already starting to shut down. Marion never regained consciousness, and she passed away about six in the morning.

Barbara and I went to the Williams Funeral Home to make arrangements and found out that Marion had already made them years earlier. They told us that she had prepaid for her funeral arrangements and that everything was taken care of. She had even prepared the obituaries that would appear in the St. Thomas and Windsor newspapers and decided on the clothes she wanted to be buried in as well as the type of luncheon to be served at St. John's Anglican Church after the funeral. She had thought of everything. I vaguely remember Marion saying something about a preplanned funeral years ago, but at the time I didn't think it was necessary. I was so wrong, but it still didn't persuade me to make arrangements for my own funeral.

Marion and I were married nearly sixty-eight years when she passed away on January 8, 2007. I felt we did well over the years, but as with other married couples, we had our ups and downs. On the whole, though, I believe we had a good relationship during our years together.

After Marion's death, I had to produce our marriage certificate for insurances, pensions, etc. I checked everywhere. I remembered we had eloped in Port Stanley, but I couldn't find any record of it. I was starting to think that maybe we had never actually gotten married. It was only after a long search that the license was finally found. It turned out that we had actually gotten married in London, Ontario, and not Port Stanley as I had originally thought. Who then got married in Port Stanley?

We all missed Marion very much. She was a wonderful woman.

Peter James McLean

Marion and I on our 65th Wedding Anniversary

The Old McLean Homestead

Whenever Barbara and Carolyn were in St. Thomas and passed by the old McLean homestead, they wanted to see the inside one more time. The outside was very well cared for even though it was now only on a half-acre piece of land. One day when they were going by, Barbara stopped the car, got out, went up to the front door, and told the owners how much we had loved the house. They were very gracious and invited us in for a tour.

It ended up there were five of us for the tour, my brother Robert, his wife, Marg, Barbara, Carolyn, and myself. It was great to see the things that stayed the same and the changes that had been made over the years.

The current owners did a wonderful job keeping the place in first-rate condition. We were able to show them where things were originally on the property such as the tennis court layout and where the four wells had been located. We even talked about the new roof on the house and what they had done with the old slate tiles they had replaced.

It was wonderful to see the house again, and we all enjoyed the time we spent there. The girls were reminded of the wonderful dinners we had there, and it brought back special memories for all of us.

Crerar

Mother and Dad never sold the property at Crerar, Ontario, that Dad had staked in 1916 and where they lived when they got married.

I remember that in the 1950s, Mother and Dad requested that I make a special trip to examine their property in Davis Township at Crerar, Ontario, in the vicinity of Sudbury. There were no roads to Crerar, so I had to walk along the railroad tracks for approximately fifteen miles to get to the location. I've never forgotten that long walk because of a bee that wouldn't leave me alone. It buzzed around my face and neck during the whole fifteen-mile trek.

The cabin was another quarter mile into the bush from the railroad tracks. The only way into or out of the property was by train. Even today there is no direct access to the property. I took a picture of the remains of the log cabin and the railroad sign that read "Crerar" to show them I had actually been there. I understand a nephew gave permission to a snowmobile club to use the property for winter trails.

After my parents died, their property at Crerar was left to my brothers, my sisters, and me. My brother Robert and I paid the taxes on the land at different times over the years, so it is still registered in the McLean name. After Robert and I are gone, we don't know what will happen to it. The property has now been in the family for over ninety-five years, and it will probably be sold for back taxes. To this day, it is still not easily accessible.

Peter James McLean

Top: Railroad tracks to Crerar
Bottom: Remains of log cabin Mother & Dad
lived in when first Married in 1911

Ted Boomer Update

Ted spent his last years as a regional manager for Capital Homes and then as general manager of Checker Cabs from 1966 until his retirement in 1979.

Ted and I often talked about life after death and what happens to us. We made a promise to contact each other if we could after we died. I never saw or heard anything from Ted after he died in 1993. He was seventy-six years old and in very poor health during his last years.

It is said that everyone dreams, but I am one of those people who do not remember their dreams. My daughters told me that since I never remember my dreams, Ted did the next best thing: he appeared to Marion in a dream. Shortly after Ted passed away, Marion said he came to her one night. He was all dressed up and was giving her one of his famous smiles. She said he looked happy and that it was like he was actually standing beside the bed. I guess, though, no one will ever know the truth until after they have died.

Soon after we became friends, Ted gave me a copy of a poem he said that he had written early in his life and I've kept a framed copy of it all these years.

WOULD THAT I COULD

Would that I could
Do some good
In this world of sorrow and strife
Would that I could
Help some soul
To lead a better life
Would that I could
Bannish sorrow
And put an end to tears
Would that I could
Hide the heartaches
And turn back the years
Would that I could
Make folks smile
And take the worry
From their face
Make them see
That life's worthwhile
Then 'twould be
A better place

Written by Ted I. Boomer
1934

Manmade Gold

Some might say that I'm just an alchemist at heart. I do believe it is possible to create gold. All it takes is the right formula from the periodic table. I have spent many years dreaming and experimenting with various formulas.

At one time I even made three bombs out of spheres I had bought at an army surplus store and packed them with my various formulas. I was hoping to create enough pressure to fuse the contents, and I set them off in one of my mineshafts. During one such experiment, my friend George Kenty, a mining engineer, was with me and he was standing close to the mouth of the shaft and got a bad cut on his arm from flying shrapnel. I had to take George to the hospital to get his arm stitched up. These experiments were far too dangerous and were a complete failure. It was back to the drawing board.

My next set of experiments dealt with new formulas. In the beginning, some looked promising but to date none have been successful. I do predict, however, that it will happen.

THE END IS NEAR

Final Move

After Marion passed away, Barbara stayed in St. Thomas to take care of me. She promised I would never be put in a nursing home, and she kept her promise. I did end up in hospital a couple of times with broken bones from falling, but I was able to return home after each hospital stay.

Barbara had six children plus grandchildren and most lived in Windsor. She wanted to move back there to be near family and friends. Barbara and I only talked about it until Carolyn was hospitalized and not expected to live. When it finally became clear that Carolyn would be okay, Barbara and I made the decision to make the move back to Windsor. Carolyn needed help and we wanted to be near all our family. Carolyn lived with her husband but her son and grandson were five hours away. Since Marion was gone, there was no reason for us to stay in St. Thomas any longer.

In 2009 we sold the house in St. Thomas and packed everything up and made the move. It was not easy emptying out my office, but it had to be done. We bought a home just two short blocks (within walking distance) from where Carolyn lived. In fact, our house numbers were the same (550).

At the time of the move I was ambulatory with the help of a walker or wheelchair, but I was slowing down. It seemed that my health was failing quite fast, and I was put under the care of hospice to be kept pain-free.

Although my "office" was gone, I still had ten mining claims in the Wawa area and had Gilbert Clement re-stake the claims as needed.

Gold fever doesn't end because you grow old, so I still had hopes of finding a large gold vein on the properties and even finding diamonds.

It was wonderful being back in Windsor with my family and good friends.

I'm Still Here?

I'm now ninety-two years old and have become completely bedridden. Most would think it was a horrible situation, right?

I felt it was freeing. I finally have time to slow down and think, to reflect on and contemplate all the things I have accomplished in my life—both the good and the bad. I know I will spend my last days at home with loving family and friends, and that eases any anxiety I have.

Wednesday nights quickly became very special. They start out with Barbara, Carolyn and friends, Nurse Mary O'Rourke-Thompson, Personal Care Assistant Lisa Valente, and sometimes Nurse Sandy Carrick is be able to join us. From the beginning the girls called it my "B, B, and B night," which they said stood for "beer, bath, and bad behavior."

The evening always begins with dinner, and then my personal nurses relieve Barbara and take over my care. While they bath and shave me, wash my hair, and do whatever else is necessary, they give me beer in my own special beer mug, and I regale them with stories. I keep trying to get them to leave their husbands and move in with me. Circumstances permitting, we have a good time. They are wonderful girls and great friends.

For someone who considered himself as independent as I always was, I took all these ministrations in stride and looked forward to Wednesday nights. It didn't stop me however, from saying to Barb every morning, "I'm still here?"

Carolyn and I finished our first attempt of this book in 2007. She kept telling me, though, that it was a "sanitized" version of my life and that I had to go back and fill in the parts I had only glossed over or even omitted. At first, I didn't think it was necessary to include my alcohol problems or my philandering in the stories. However, Carolyn finally convinced me that the whole truth needed to be told, and I have tried to give Carolyn all the information she needs to finish the book.

I realize I won't be here much longer, but Carolyn has promised that the book will be finished and published. I will take her at her word.

Lisa Valente, Mary O'Rourke-Thompson, Daughters Carolyn, Barbara and Sandy Carrick

EPILOGUE

The Wednesday night dinners continued for some time after Dad passed away on June 26, 2010. Dad was missed by everyone.

Then, unexpectedly, Barbara passed away on January 30, 2011. It was a great shock to all of us. She was well loved and respected by everyone who knew her. Barbara was a wonderful person and a caring nurse, and she was always there for her family and friends. She was a wonderful mother, grandmother, great-grandmother, daughter, sister and aunt. She is greatly missed by all who knew and loved her. Over the years, Barbara was my own personal nurse and best friend.

The railroad station in St. Thomas, Ontario, where both my Grandfather James Guy McLean and my Father Peter James McLean started working all those years ago is being refurbished into a Museum and the North American Railway Hall of Fame. My brother Guy had a brass plaque engraved in their honor and had it attached to a door he purchased. The door leads to the ticket master's office, main floor passenger entrance, which is now used as the main office for the Railway Hall of Fame.

On June 6, 2014 Dad was inducted into the North America Railway Hall of Fame for his Technical Innovations. The family is proud and he would have been pleased to have been recognized.

Carolyn (McLean) Bosetti

North America Railway Hall of Fame

ST. THOMAS, ONTARIO, CANADA

2014

Hall of Fame Induction Award

The Board of Directors on the recommendation of the Selection Committee hereby inducts

Peter James McLean

in the category of

**Technical Innovations
(Local)**

*to the
North America Railway Hall of Fame*

*Given at St. Thomas Ontario, Canada
on the 6th day of June, 2014.*

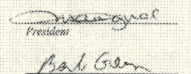

President
Secretary

North American Railroad Hall of Fame Induction Award

TIMELINE REFERENCE

1930s
> Started as an apprentice journeyman machinist with the New York Central Railroad.
> Continuously employed by railroad until retirement in 1980.
> Built numerous scale models of steam engines showing different valve options.

1940s
> Developed and patented magnetic water softening devices for feed water in boilers of steam locomotives (no scale buildup).
> Advanced to steam engine and diesel supervisor for New York Central, Penn-Central and Conn-Rail Railroads.
> Took steam - and diesel-engine-associated courses through International Correspondent School.
> Took a Dale Carnegie course on effective speaking.
> Invented and patented a large range of toy guns.
> Served one term as an alderman on St. Thomas City Council and was appointed chairman of the health department.
> Invented and patented a cigarette holder that filtered the smoke through water.

1950s
> Invented and patented the underwater Waterscope.
> Owned US icebreaker *SS Chaparral* to explore the Great Lakes for sunken treasure.
> Owned and operated an underwater diving school.

Researched and published maps of sunken ships in the Great Lakes.

Started to prospect and develop mining properties for uranium, gold, silver, graphite, moylendite, garnets, precious stones, black granite, amber, mica, and corundum in all areas of Ontario.

Grew synthetic gems and did synthetic diamond experiments in cooperation with the National Science Research Council of Canada in Ottawa.

1960s

Invented and patented Gas-O-Rator. Was first to use exhaust liquids and fumes vented back to gas engine for recycling and improved gas mileage. Ran tests for Gas-O-Rator on a 1955 high-mileage Cadillac.

At the same time, invented and patented an oil purification system.

Continued prospecting and developing mining properties.

Researched and published *Treasure Map of Canada*, showing lost gold mines, buried treasures, and large gold and silver mining areas in Canada.

1970s

Continued prospecting and developing mining properties.

Invented, developed, and patented high-production and efficient water distilling equipment for retail and wholesale markets by using the condensation of steam to preheat the boiler feed water.

Researched, compiled information and data, and wrote a book, *Documentary Gold*.

1980s

Continued prospecting and developing mining properties.

Organized prospecting classes held in the main branch of the Windsor Public Library.

1990s

Continued prospecting and developing mining properties.

Researched, compiled information, and wrote *Mining and Minerals*.

Researched and compiled information and data to write and publish book on prospecting in Canada (for fun and profit).

2000s
- Became the president of National Gold and Research Development Inc.
- Continued to research and compile information and data to write and publish books.
- Owned gold mine in Northern Canada and used its deep underground shafts for explosive and dangerous portions of experiments.
- Owned large research library and files on diamonds in Ontario.
- Owned over one thousand geological reports, maps, and mineral deposit information for all areas of Ontario.
- Owned large collection of all types of minerals found in Ontario.
- Owned large amount of mining equipment and prospecting supplies.
- Owned gold dredges, sluice boxes, etc.
- Owned lab equipment for testing mineral content of ore..
- Owned approximately 520 acres of mining property along the Michipicoten River in Wawa, Ontario, and dredged for gold and diamonds.

Focus of Work

To make information available about the great wealth, riches and treasures of undiscovered Natural Resources in Canada. Ninety-five percent of all the large mines in Canada have been discovered first by prospectors before being taken over by the big companies. The majority of Northern Canada is undeveloped, unexplored and open for future prospecting. Continue with experiments to manufacture pure gold.

Printed in the United States
By Bookmasters